Dancing with Water

The New Science of Water

Dancing with Water

The New Science of Water

A Guide to Naturally Treating, Structuring, Enhancing, and Revitalizing Your Water

Second Edition

By MJ Pangman & Melanie Evans
www.dancingwithwater.com

Uplifting Press

Dancing with Water
The New Science of Water

A Guide to Naturally Treating, Structuring, Enhancing, and Revitalizing Your Water
Second Edition

www.dancingwithwater.com

Photography: Michael Martin and others as noted in image credits at the end of each chapter

Cover Design & Illustrations: Alicia Pangman
Copy Editing: Erin Cler

Library of Congress Control Number: 2016914022
ISBN: 978-0-9752726-3-3
Printed in the United States of America

Praise for *Dancing with Water*

"This book is a must read for everyone interested both in modern science and in their own health. We are coming to a new understanding of water, not just as a source of vitality and life, but as life itself. This vision is supported by a new set of scientific data which we find referenced in the book. What is equally important is the fact that it is written in simple, understandable language that allows the reader to feel the flavor of the modern research atmosphere. Another important aspect of the book is found in its practical applications. *Dancing with Water* is a very important and useful guide both for professionals and all interested people." —Konstantin Korotkov, PhD, professor at Saint-Petersburg University of Informational Technology, Mechanics and Optics. www.korotkov.org

"In many ways, the new science of water is not just about scientific progress, it is also about the evolution of important concepts and a new philosophy of science that allows new concepts to become mature scientific platforms able to surpass outdated theories. In the second edition of *Dancing with Water*, the authors successfully present important discoveries from the frontiers of modern science revealing how water's interaction with natural forces can change its structural geometry. Written in simple and understandable language, this book is highly recommended for those interested in the emerging new science of water." — Adam B. Dorfman, Founder of *Concept Evolution* magazine http://conceptualrevolutions.com/

"The new renaissance in our understanding of water is guiding us to radically rethink our outdated perceptions of water's role in the physical body, and in fact, throughout the entire cosmos. *Dancing with Water* pioneers this renaissance, emphasizing our existence as water-based beings and our connection with the cosmos through water's living matrix. Not only does the book provide the background

for understanding this elegant and mysterious substance, it teaches us how to honor and respect it as we return its natural, life-giving properties. The first edition of *Dancing with Water* has been a companion of mine for years. It has been the first book I recommend to anyone for gaining an overview of water's essence–everything from the latest science to the spiritual aspects of this essential substance. Now, with the second edition, the authors update the science and deepen their commitment to help each reader develop a personal relationship with water, and to enter into the *dance of life* with a new perspective. A fabulous read for scientists, health practitioners, and anyone interested in the true nature of water and themselves."
—Matt Thornton, Director; Emoto Peace Project, United Kingdom and Ireland.

"*Dancing with Water* reveals both the sacredness and the complexity of water in a unique blend of science and spirit where the authors engage the reader in the ultimate "Dance with Water." From life's beginnings in the ocean, to the geometries that sustain life, learn how water dances with all the elements and the fundamental forces of nature. Written for the layperson, yet documented for the scientific community, this book is a must if you are looking to gain insight into the mysteries and therapeutic properties of water."
—Karen Elkins, publisher, *Science to Sage* magazine

Dedication

Dancing with Water is dedicated to the pioneers of the past, whose writings, and work with water have shaped our thinking and guided the unfolding of this book. We acknowledge:

- Viktor Schauberger—Austrian forester and champion of Nature
- Albert Szent-Györgyi—Nobel prize laureate and father of modern biochemistry, remembered for his ability to think outside the box
- Mu Shik Jhon—honored for his 40 years of research on the molecular structure of water
- Rudolf Steiner—developer of biodynamic agriculture, Waldorf education, and anthroposophical medicine
- Slim Spurling—developer of the Light-Life (Tensor Ring) technology, and dedicated to a pollution-free planet
- Mae-Wan Ho—cofounder of the Institute of Science in Society, whose understanding of quantum coherence has opened new vistas to our vision of water

Dancing with Water is also dedicated to the pioneers of the future who *are developing* new ways to renew the water on the planet.

Lastly, this book is dedicated to our partner in this endeavor—Water—the key to life and the carrier of the life force for the inhabitants of Mother Earth.

CONTENTS

Preface

Many books have been written about the importance of water for a healthy body and mind. Even more has been written about contaminants, filtration, and purification. Everywhere we read of diminishing water resources, water rationing, and the privatization of water. All this is evidence that we are becoming more aware of the significance of our relationship with water. Yet, so far, little emphasis has been placed on understanding water as a living essence. When we learn to treat water as a conscious participant in the process of life, rather than as a commodity to be harvested, sold, and abused, we will begin our return to the natural order of things. Only then will we enter into what Charles Eisenstein refers to as the "Age of Water."

Eisenstein identifies the *Age of Water* as an era during which we instinctively treat the Earth and everything on it as sacred. This long-awaited time is at our doorstep now. It is a time to clasp hands with Mother Nature and return to the garden into which we were born. It is a time to discover the keys to creation (the same keys that will create sustainable solutions for our future). As we step into the Age of Water, we will discover that returning the *life force* to water is synonymous with stepping into *life* in a whole new way.

New science reveals that *organization* is as important as *ingredients* when it comes to life. This revelation provides new significance as we look at water's *liquid crystalline* geometry. Coherent molecular organization establishes a superior medium for the transmission of signals and information. According to Mae-Wan Ho, "Liquid crystalline water is involved in the relay and storage of information; in the amplification of biological signals; and in the transduction of a variety of forms of energy." Evidence of molecular organization also establishes firm ground for the acceptance of water as a *living* component of our world.

Dancing with Water will help you to personally step into the Age of Water. It will teach you how to return its liquid crystalline order, its vitality, and its ability to support life to the fullest. As a guide to naturally treating, structuring, enhancing, and revitalizing your water, *Dancing with Water* will also help you to connect with water on a conscious level and to honor the role water plays in the dance of life. The authors' backgrounds (one in science, the other in the healing and intuitive arts) offer a unique perspective. The book combines existing science with wisdom available through the intuitive arts—balancing and expanding on the best science has to offer.

Part I of *Dancing with Water* introduces water's unique qualities and the interplay of natural forces that influence the development of water's liquid crystalline state. The chapters in Part I discuss movement and vortices, minerals and gases, electromagnetic forces, the effects of light and sound, the flow of energy, and the organizing influence of geometry.

Part II of the book is a *dance guide* with step-by-step instructions for how to use Nature's tools to create liquid crystalline water capable of carrying signals and messages deep into the tissues and cells of your body. Practicing the "dance steps" in Part II will bring a new level of awareness (and a new level of health) based on superior hydration, more efficient detoxification, improved cellular communication, and enhanced nutrient absorption. These benefits translate to improved energy, softer skin, faster recovery, graceful aging, heightened awareness, and the potential reduction of a variety of symptoms.

The authors invite you to step in with them now . . . and immerse yourself in the *Dance with Water*.

Part I

Getting to Know Your Dance Partner

Introduction to Part I

The only way to truly dance is to be in sync with your partner—to anticipate each other's moves and to generate a flow of energy that supports and carries the dance. Dancers who are able to connect in this way literally *ride the energy* and create the dance as they go.

One of the biggest challenges encountered by dancers is getting to know a new partner. It requires a desire to understand each other at a level that supports the exchange of unspoken communication. This degree of connectedness inspires the intertwining of energy such that the skills of each partner are enhanced by the other. As your partner in life, water must be understood on a new and deeper level if you are to be able to consciously engage in the dance of life. Doing so will enable you to rise to new levels of awareness and to discover your full potential.

Part I of *Dancing with Water* is designed to help you get to know water as your partner in life. It includes gaining an appreciation of the way water moves, becoming aware of the things that help water to stay balanced, recognizing ways to enhance her energetic potential and becoming aware of how she carries the music to sustain the dance.

Chapter 1

The New Sciences and the Liquid Crystalline Structure of Water

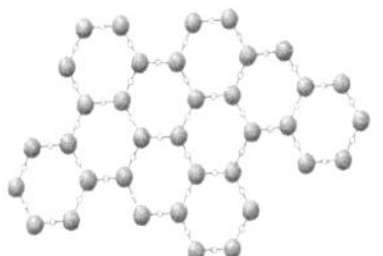

Despite the fact that the scientific community has amassed libraries of information on water, our current knowledge of this life-supporting liquid is still trivial—a drop in the ocean. Yet the *new* sciences are opening a refreshing new chapter to the book on water, allowing us to peer beyond many former limitations.

The *old* sciences speak to our intellect. They function inside a carefully crafted box encumbered by the slow, linear progression from hypothesis to acceptance of new ideas. Although there is nothing wrong with this approach, it tends to dismiss ideas that cannot be tested using traditional methods from inside the box. The old sciences limit us to what we can experience with our senses. They teach us that reality is defined by what we can see, hear, and touch—forgetting that our senses themselves are limited. The old sciences also isolate and compartmentalize every area of biology, taking things apart to discover how each part works. Unfortunately, in doing so, the big picture has faded into the background. At the same time, we have lost sight of how each part contributes to the wholeness of life itself.

The *new* sciences speak to a heart-centered intellect. They take us to the subatomic world, and yet they reveal how all things are a part of the whole. The new sciences show us the inadequacy of our physical senses for evaluating the universe. They honor both the intuitive and the physical senses, bridging the gap between the right and left brain as well as the gap between science and spirit. The new sciences often reveal that nothing changes without affecting everything else. From this new perspective, there are no boxes, there is no separation, and we are all participants in the ongoing creative process.

Where water is concerned, the new sciences reveal a fourth phase of matter—a liquid crystal designed to interact with our biology. When used consciously, liquid crystalline water can heighten our experience as water beings on the blue planet we call home.

Crystals and liquid crystals

The old sciences teach us that there are three phases of matter: solid, liquid, and gas. The new sciences reveal a fourth state of matter, referred to as a liquid crystal. In this state, molecules form repeating geometric arrays that have fluidity. The resulting substances have qualities of liquids and *crystals.*

Crystals are unique solids. They are well known in the mineral kingdom for their angular shape and often for their visual clarity. But crystals have other interesting properties that make them ideal for sound and light amplification (in radios, TVs, and lasers) and as frequency control devices (in clocks, computers, and navigation equipment). As semiconductors, crystals have the capacity for processing information. They can detect, switch, store, modulate, filter, rectify, and amplify energy.[1] Liquid crystals (including water) exhibit many of these same properties.

By definition, crystals are solids whose atoms or molecules are arranged in an ordered pattern that extends in three spatial directions. Each mineral has a specific geometric pattern determined by the way its molecules form their repeating crystalline arrangement.

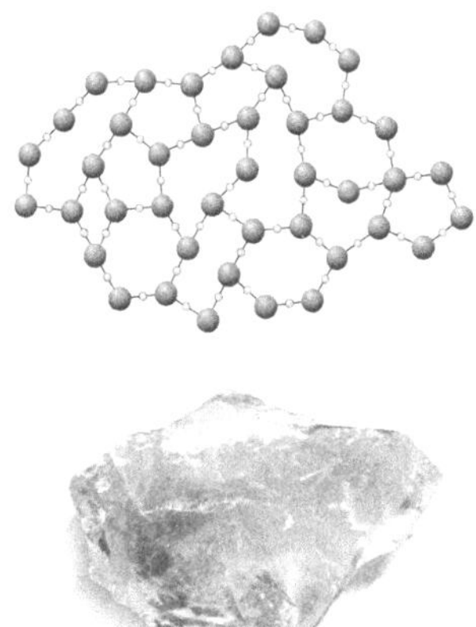

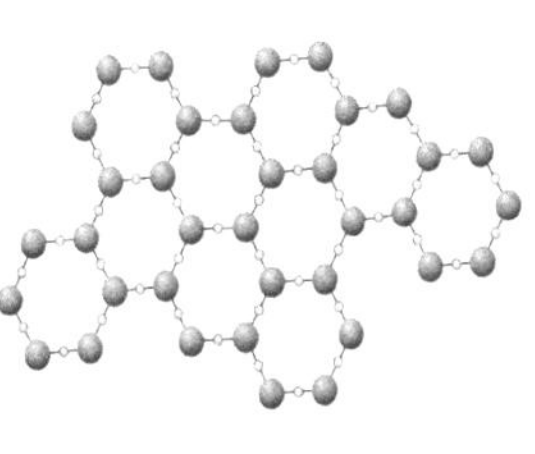

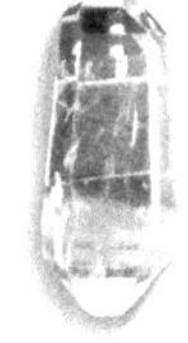

The difference between a liquid and a *liquid crystal* is similar to the difference between a solid and its corresponding crystalline form. For example, quartz (silicon dioxide [SiO_2]) exists everywhere on Earth in an amorphous form that most people would call a quartz rock (above left). On a molecular level, the SiO_2 molecules in a quartz rock are random, as shown in the above illustration. However, under certain conditions, SiO_2 molecules organize to form a repeating geometric pattern known as a quartz crystal (above right). Although both materials are made of the same SiO_2 molecules, organization changes their *physical appearance* and their *properties*. The same is true of a diamond. Lacking organization, it is simply a chunk of coal. However, under heat and pressure, carbon atoms become organized. Coal and diamonds, like quartz and quartz crystals, are made of the same original elements. Organization changes their properties.

Liquid crystalline water

Liquid crystals are a special phase of matter in which liquids retain their fluidity while assuming an ordered state. Molecules within a liquid crystalline arrangement are held together by electrostatic forces and supported by each other much like bricks in a 3-D structure. Yet, in the case of liquid crystals, the structure retains flexibility: when one molecule moves, they are more likely to move together. In liquid crystalline water, organized molecular sheets may slide past one another while retaining organization within the whole.

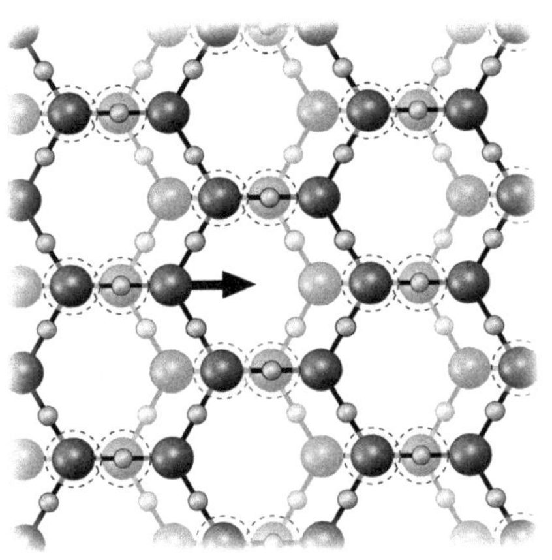

Dr. Gerald Pollack, professor of bioengineering at the University of Washington, has clarified the most likely geometric configuration for liquid crystalline water (which he refers to as EZ water to describe the exclusion zone that the liquid crystalline state creates). According to Pollack, liquid crystalline water forms stacks of honeycomb-like sheets that are offset to maximize cohesive forces (upper right graphic). This arrangement offers a variety of structural variants whereby successive sheets can shift to any of the six strut directions to accommodate irregularities around water molecules.[2] (lower right graphic). Honeycomb-like sheets are composed of linked hexagonal rings. For this reason, liquid crystalline water has often been referred to as hexagonal water.[3]

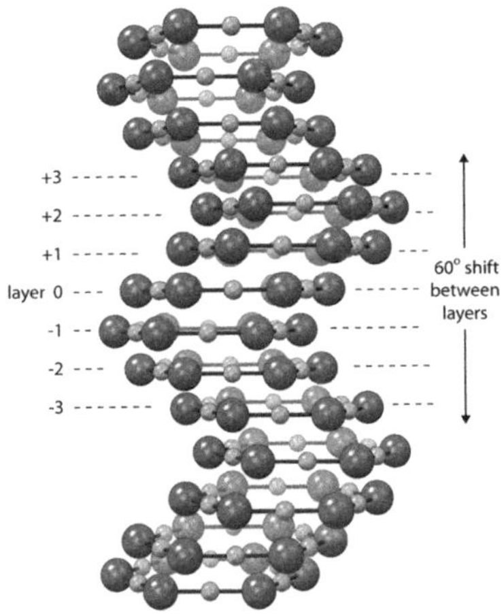

The next time you sit at your computer or in front of a video display terminal, notice how pressing on the display screen produces obvious fluidic movement just beneath the surface. This is a liquid crystal—the heart of LCD (liquid crystal display) technology. An LCD is made by sandwiching a liquid crystalline layer between two electrodes. Depending on the amount of voltage sent through the electrodes, the molecules of the liquid crystal shift to form colors and shades elucidated with the use of polarizing filters. Without the liquid crystal, the display would not work. Liquid crystals are well known and their properties are well understood. Like solid crystals, liquid crystals provide an efficient mechanism for storing and transferring energy. Yet liquid crystals have significantly greater versatility than solid crystals. They are flexible and many times more responsive.

Hydrogen bonding

Water's liquid crystalline structure is made possible, in part, by electrostatic forces called hydrogen bonds. In the water molecule, oxygen atoms maintain a slightly negative charge and hydrogen atoms maintain a slightly positive charge. These charges join water molecules together to form the interconnected liquid crystalline network.

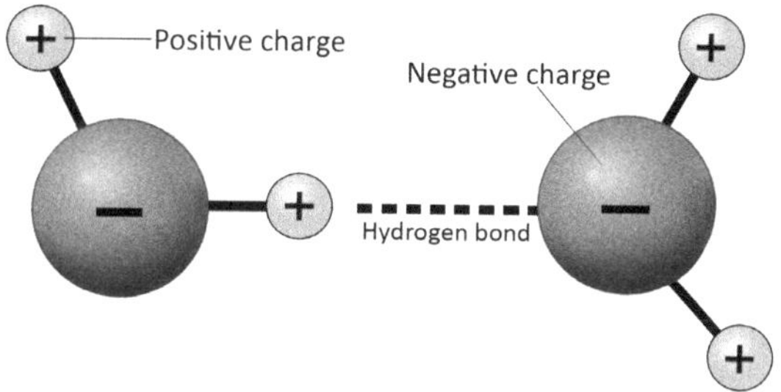

The Hydrogen Bond

The positively charged hydrogen atom from one water molecule is electrostatically attracted to the negatively charged oxygen atom from another water molecule. This bond is the foundation for the development of the liquid crystalline water matrix.

In a large portion of the water on the planet, hydrogen bonding is random. In this water, hydrogen bonds break and reform billions of times a second. However, Nature provides forces that can influence hydrogen bonding. Under some circumstances, water molecules become organized and develop a liquid crystalline matrix in which organization itself helps stabilize the hydrogen bonds.[4,5]

Liquid crystalline water (also called *structured* or *ordered* water) is a matrix within which energy can be stored and through which information may be easily delivered. The greater the degree of structure (the more organized the matrix, characterized by increased hydrogen bonding), the more efficiently water can store and deliver information. In other words, water's capacity to act as a crystal is directly proportional to the degree of hydrogen bonding. Its ability to act as a crystal is also dependent on coherence.

Coherence

Coherence is defined as a *fixed phase relationship* between waves or particles whereby the movement of one wave/particle coincides with the movement of all of them. For a system to exhibit coherence it must operate as though it is one unit—even though it is composed of multiple, independent parts. The healthy human body is a good example of a coherent system. It comprises individual cells, tissues, and organs, yet it functions as one unit. Every cell is like a separate musical instrument; every organ is like a section of the orchestra. Together, they create a living symphony. When it is healthy, the human body is coherent on a quantum level, where everything, from the tiniest subatomic particles to larger organs and tissues, functions as a whole. With this level of cooperation, the symphony of life, referred to as *quantum jazz* by Dr. Mae-Wan Ho, is flawlessly composed and played at the same moment.[6] This is *quantum coherence.*

Liquid crystalline water is a structured, coherent, fluid crystal in which individual molecules cooperate and function as though they are one—like a piece of fabric blowing in the wind. The fabric is created by hydrogen bonds that function as *threads* holding water molecules together. Coherence enables each molecule to communicate with every other molecule in the system—instantaneously.

As long as a system maintains coherence, it also maintains organization.[7] In many ways, it is the *flow of energy* through a system that creates and sustains coherence. In her seminal work, *The Rainbow and the Worm: The Physics of Organisms*, Mae-Wan Ho demonstrates how the flow of energy through an organized system helps to create and support the structural organization of the system, and likewise, how the structure of the system supports the flow of energy.[8] Living organisms depend to a large extent on the coherence of their internal water matrix.[9]

The new sciences have embraced higher ways of defining life itself. Rather than focusing on chemistry or on the presence of specific elements, recent theories revolve around organization and coherence to describe the difference between living and nonliving matter.[10,11] Living matter also displays a strong electromagnetic field, sometimes referred to as an aura.

Life fields

Harold Burr, a Yale University School of Medicine neuroanatomist, theorized in the 1930s that energy fields were the "organizing principle" that kept living tissue from falling into chaos.[12] Burr was one of the first scientists to measure the electromagnetic field around embryos, plants, and animals. He theorized that characteristics of an organism's "field" provided clues to the health and wholeness of the organism. Valerie Hunt authenticated Burr's work by substantiating the existence of a dynamic bioenergetic field around all living things.[13] Her work described the human bioenergy field as being composed of "balanced, coherent energy patterns flowing across the full spectrum of frequencies."[14] Hunt's research revealed that chaos, or anticoherence, in the energy field corresponded with ill health and disease.

Today, Dr. Konstantin Korotkov, at the Saint-Petersburg University of Informational Technology, Mechanics and Optics, has developed a technology to map and quantify the life field around the human body and other living matter—including water.[15] His technology (featured in Appendix A) substantiates the dynamic (ever-changing) nature of water's energy field and is able to quantify these changes, which are influenced by water's structure and coherence. As with biological organisms, organization and coherence at the physical level are mirrored by coherence in the energetic field.

Crystals as information carriers

Water has the capacity to store and transfer information. This is one of its pivotal biological roles. This capacity accounts for the instantaneous transfer of signals and other vibratory information within biological organisms. Liquid crystalline order is the key. Energy and information may be stored as vibrations in molecular bonds and within the structure of the system.[16,17] The ordered crystalline matrix provides a pathway for the efficient transfer of information.

The field of solid state physics focuses on highly ordered systems with an emphasis on the unique properties that emerge when atoms and molecules come together in crystalline order. One of these properties concerns the sharing of electrons in a crystalline system. Albert Szent-Györgyi (1893–1986), the father of modern biochemistry, commented on this relationship as early as 1941:

> If a great number of atoms be arranged with regularity in close proximity, as for example in a crystal lattice, single valency electrons cease to belong to one or two atoms only, and belong instead to the whole system. A great number of molecules may join to form energy continua, along which energy . . . may travel.[18]

Dr. Szent-Györgyi was one of the first to recognize that crystalline order creates an *energy continuum* as described above. This feature of liquid crystalline water (delocalized/free electrons) is characterized by the absorption of UV light at the 270 nm wavelength.[19] It is commonly noted with hexagonal "ring" structures at the molecular level. Today, we know that more than electrons participate in the flow of information through the crystalline continuum. James Oschman, author of *Energy Medicine*, describes other energy units that can convey information within the crystalline continuum. Besides electrons, they include protons, phonons, plasmons, holes, excitons, and solitons, among others.[20]

Much of the water in the healthy human body is in a liquid crystalline state. In fact, the degree to which the water in the body is organized/structured has been correlated with age[21] and health.[22] The new sciences reveal that many components of the body are liquid crystals themselves, including connective tissue (collagen), muscle tissue, nerve sheaths, cell membranes, and DNA.[23] We have been aware for some time that organized water surrounds healthy DNA. This water is responsible for the DNA's stability,[24] and as water loses its crystalline structure, the integrity of the DNA may be compromised.[25]

Liquid crystalline components of the body work cooperatively with liquid crystalline water to create an information network that reaches to every cell.[26] It has been hypothesized that even acupuncture meridians (channels that energetically link every part of the body) are chains of liquid crystalline water.[27,28] Valerie Hunt demonstrated that acupuncture meridians and connective tissue function together as a channel between the living organism and its energetic field.[29] From this perspective, water is not only the key to life; it may also be our link with everything in the universe.

Factors that affect water's molecular structure

Nature provides many ways to support water's liquid crystalline state. Outside of biological organisms, these ways include everything from vortices and minerals to organizing forces such as light, sound, magnetism, and piezoelectricity. They include the influence of turbulance, temperature, and the geometry inherent in the universe itself. Viktor Schauberger understood these natural forces like no other individual—then or now.

Viktor Schauberger (1885–1958) was one of the most profound thinkers of the 20th century. Though unschooled in traditional science, he developed an intuitive wisdom as he observed Nature in the Austrian forest. His understanding of the energetic forces in and around water became so broad that he accomplished feats that are still unsurpassed today. We are reminded of his credo, *Comprehend and emulate Nature*, as we seek to develop a more balanced and sustainable way of life. According to Schauberger, enhancing water's natural properties for

Viktor Schauberger

the improvement of our health and the well-being of the Earth is not necessarily difficult. With a basic understanding of the forces that influence water's structure, anyone can emulate Nature.

Movement/turbulence

Think of the words that have been used to describe water: *splashing*, *bubbling*, *gurgling*, *whirling*, *spinning*, and of course, *dancing*. Each of these words implies not just movement but also freedom to interact with the environment. Freedom of movement is as important for water as it is for us. When it remains stagnant for too long, water can accumulate microbes and wastes—just as we do when we neglect exercise. For water, movement provides the opportunity to breathe (exchange oxygen and other gases). It gathers energy as it circulates and reverberates off obstacles in its path.

When we talk about movement, we are not referring to forced (pressurized) movement through pipelines. Pressurized water delivered through straight pipes acquires static-like energy. Water does not have to travel far before its structure is compromised and its energy depleted. In this state, it is no longer able to carry the finely tuned information necessary to sustain high-quality life. When water arrives in homes and places of business via pressurized delivery systems it has acquired an aggressive nature that *robs* rather than *supplies* life force. This was illustrated by Schauberger and Professor Franz Pöpel, who conducted a series of experiments at the Technical College in Stuttgart, Germany. One of their most intriguing discoveries revealed that water flowing through a straight pipe experienced *more* resistance than water flowing through a spiral enclosure. Equally surprising was their discovery that water flowing through a spiral enclosure experienced less resistance as the velocity increased. At certain velocities, the resistance dropped below zero.[30] In other words, water that is delivered in pressurized pipes, such as those in use today, encounters extreme resistance. This resistance increases the amount of force required during delivery and at the same time strips water of its crystalline structure and vital energy. Water that is allowed to move freely stays more energized than water that is pressurized because of the spiraling vortices water produces as it moves with freedom.

Spirals and vortices

There are no truly straight lines in the natural world. In fact, if Nature had a predominant form, it would be the circle or the ellipse. A circle gives rise to the spiral—Nature's way of gathering energy and organizing matter.

Water naturally spins. Each time it encounters an obstacle, the forward motion is curtailed because water curls inward, forming a spiral. Spiraling movement is one of the ways Nature gathers energy. A meandering river provides a good example. As water travels down a riverbed, the central axis becomes a huge horizontal vortex interacting with smaller vortices around it. At each bend in the river, the vortex reverses direction. In this way, water gathers and dispenses energy throughout its journey. Even though water often takes a *longer* path, it always arrives with a greater degree of energy than if it had been forced to follow a shorter, straight course.[31]

The spin of a vortex produces powerful forces (think of a tornado or hurricane). Inside layers flow much faster than outer layers. Theoretically, the velocity at the center can approach infinite speed. This explains another phenomenon discovered by Schauberger and Pöpel: vortices cause molecules to become organized. When fine particles were placed into spiraling water during their experiments, the particles were pressed together so tightly that they formed extremely dense masses.[32] Much of Schauberger's work revolved around the use of vortices. Chapter 3 is devoted to a discussion of vortices and their impact on water.

Electromagnetic fields

There is no doubt that electromagnetic fields influence the structure of water. Water is a polar molecule with a positively charged and negatively charged side. The Earth is bathed in electromagnetic fields—everything from its own magnetic field to the electrical charges that make up the ionosphere. The sun showers a full spectrum of electromagnetic wavelengths on the Earth, while rocks and minerals emit subtle piezoelectric influences of their own. Everywhere, water is influenced by electromagnetic fields, some of which have an organizing effect and others that have a disruptive effect. Chapter 9 discusses the influences of magnetism and electricity on water's liquid crystalline geometry. Other chapters outline the influences of light, vibration, geometry, and stillness on water's developing structure.

Temperature

Temperature also influences water's structure. As the temperature decreases, hydrogen bonding increases.[33] Water is most dense (most tightly packed and able to support the greatest weight) at a temperature of 4° C (39.2° F). This temperature is referred to as the *anomaly point* of water. Other factors aside, water is more structured at 4° C than at any other temperature; it is at its energetic peak. Schauberger utilized this knowledge in the building of giant logging flumes that carried logs through the forest to the sawmill. By periodically adding an influx of colder water, he was able to support the larger logs that would have ordinarily sunk and been caught in the twists and turns of the flume.[34] Keeping water cool contributes to its structural organization as well as to its capacity to hold gases.

Minerals and gases

Minerals and gases play a dynamic role in helping water to maintain structure, purity, and energy. Minerals act as nucleating zones for the initiation of more complex liquid crystalline order. As receptors, minerals also provide the basis for a vibrant exchange of energy. Gases, particularly hydrogen, oxygen, and carbon dioxide, help water to stay clean, energized, and balanced. They provide the same service for the body when they are present in water. Chapter 4 is a discussion of the importance of minerals in liquid crystalline water. Chapter 5 is a discussion of the role of gases.

Full-spectrum living water

When water, supplied with minerals and gases, is allowed freedom of movement in the presence of natural organizing fields (magnetism, piezoelectricity, light, sound, and others), it forms a liquid crystal. Endowed with vibratory information found in Nature, it becomes *full-spectrum living water*—a coherent, living, liquid crystal that is capable of supporting organic life to the fullest.

Hundreds of years of preparation within the Earth can be nullified in a few minutes when we lack understanding of how to emulate Nature. Water that emerges from springs as *full-spectrum living water* can be rapidly stripped of its energy as it travels through pressurized pipes or during treatment at municipal facilities, where further atrocities are committed. The unconscious use of water for dispensing agricultural chemicals and disposing of industrial waste is a desecration of water on the planet.

Fortunately, many are awakening to the realization that liquid crystalline structure and life force can be returned to water—often as easily as they have been taken. By emulating Nature, as Viktor Schauberger and others have admonished, we can revitalize not just the water we drink but also the water on the entire planet. The remaining chapters will help you to understand the forces in Nature that contribute to water's life force. They will show you how to reestablish water's liquid crystalline structure and how to add *information* to maximize its potential.

Image credits (Chapter 1)

Page 4 EZ honeycomb sheets and side view:
Pollack, G. (2013) *The Fourth Phase of Water,* Ebner & Sons.

Page 9 Viktor Schauberger: Schauberger-Archive, Germany, Bad Ischl

References (Chapter 1)

1 Oschman, J. (2003). *Energy Medicine in Therapeutics and Human Performance.* Butterworth-Heinemann, p. 93.

2 Pollack, G. (2013). *The Fourth Phase of Water: Beyond Solid, Liquid, Vapor.* Ebner & Sons, pp. 59-60.

3 Jhon, M-S. (2004). *The Water Puzzle and the Hexagonal Key*. Uplifting Press, pp. 58-61.

4 Luck, W. (1998). The Importance of Cooperativity for the Properties of Liquid Water. *Journal of Molecular. Structure*, 448(2), pp. 131-142.

5 Pollack, G. (2013). *The Fourth Phase of Water: Beyond Solid, Liquid, Vapor*, Ebner & Sons, pp. 66-68.

6 Ho, M-W. (2008). Quantum Coherent Liquid Crystalline Organism. Invited lecture at European Quantum Energy Medicine Conference, Copenhagen. Available online: http://www.i-sis.org.uk/QuantumCoherentOrganism.php.

7 Del Giudice, E. and Tedeschi, A. (2009). Water and the Autocatalysis in Living Matter. *Electromagnetic Biology and Medicine*, 28 (1), pp. 46-52.

8 Ho, M-W. (1998). *The Rainbow and the Worm: The Physics of Organisms*. 2nd ed. World Scientific, p. 75.

9 Ho, M-W. (2009). Organism as Polyphasic Liquid Crystalline Water. Invited lecture, Dept. of Mechanical Engineering, University of Belgrade Physical Medicine and Serbian Medical Association, Belgrade, Serbia. Available online: http://www.i-sis.org.uk/isisLectureQuantumMedicine.php

10 Ho, M-W. (2008). *The Rainbow and the Worm: The Physics of Organisms*. 2nd ed. World Scientific, pp. 170-171.

11 Pollack, G. Figueroa, X. and Zhao, Q. (2009). Molecules, Water, and Radiant Energy: New Clues for the Origin of Life, *International Journal of Molecular Science* 10 (4), pp. 1419–1429. Available online: http://www.ncbi.nlm.nih.gov/pmc/articles/PMC2680624/

12 Burr, H. (1972). *The Fields of Life*. Ballantine Books.

13 Hunt, V. (1996). *Infinite Mind: Science of the Human Vibrations of Consciousness*. Malibu Publishing.

14 Hunt, V. (1996). *Infinite Mind: Science of the Human Vibrations of Consciousness*. Malibu Publishing, pp. 328-348.

15 Korotkov K. and Krizhanovsky, EV. (2004). Time Dynamics of the Gas Discharge Around Drops of Liquids, in: *Measuring Energy Fields: State of the Science*, Korotkov, K. (ed.), Backbone Publishing, pp. 103-123.

16 Ho, M-W. (1998). *The Rainbow and the Worm: The Physics of Organisms*. 2nd ed. World Scientific Publishing, p. 77.

17 Oschman, J. (2003). *Energy Medicine in Therapeutics and Human Performance*. Butterworth-Heinemann, p. 297.

18 Szent-Györgyi, A. (1941). The Study of Energy Levels in Biochemistry. *Nature*, 148 (3745), pp. 157-159.

19 Pollack, G. (2013). *The Fourth Phase of Water: Beyond Solid, Liquid, Vapor.* Ebner & Sons, p. 58.

20 Oschman, J. (2003). *Energy Medicine in Therapeutics and Human Performance.* Butterworth-Heinemann, pp. 89 and 248-254.

21 Katayama, S. (1992). Aging Mechanism Associated with a Function of Biowater. *Physiological Chemistry and Physics and Medical NMR*, 24, pp. 43-50.

22 Damadian, R. (1971). Tumor Detection by Nuclear Magnetic Resonance. *Science*, 171 (3976), pp. 1151-1153.

23 Ho, M-W. (1998). *The Rainbow and the Worm: The Physics of Organisms.* 2nd ed. World Scientific, p. 173.

24 Pal, SK., Zhao, L., and Zewail, AH. (2003). Water at DNA Surfaces: Ultrafast Dynamics in Minor Groove Recognition. *Proceedings of the National Academy of Sciences*, 100 (14), pp. 8113-8118.

25 Jhon, MS. (2004). *The Water Puzzle and the Hexagonal Key.* Uplifting Press, pp. 58-61.

26 Pischinger, A. (2007). *The Extracellular Matrix and Ground Regulation.* North Atlantic Books.

27 Lo, SY. (2004). *The Biophysics Basis for Acupuncture and Health.* Dragon Eye Press.

28 Ho, M-W. and Knight, DM. (1998) Liquid Crystalline Meridians. *American Journal of Chinese Medicine*, 26, pp. 251-263. Available online: http://www.i-sis.org.uk/onlinestore/papers1.php#section3

29 Hunt, V. (1996). *Infinite Mind: Science of the Human Vibrations of Consciousness.* Malibu Publishing, p. 242.

30 Alexandersson, O. (1990). *Living Water: Viktor Schauberger and the Secrets of Natural Energy.* Gateway, pp. 118-119.

31 Schauberger, V. (1998). *The Water Wizard* (translated and edited by Callum Coats). Gateway, pp. 29-33.

32 Alexanderson, O. (1990). *Living Water: Viktor Schauberger and the Secrets of Natural Energy.* Gateway, p. 117.

33 Chaplin, M. Water Structure and Science website: http://www1.lsbu.ac.uk/water/explan.html

34 Alexanderson, O. (1990). *Living Water: Viktor Schauberger and the Secrets of Natural Energy.* Gateway, p. 25.

Chapter 2

Balance, Polarity, and Cycles

Full-Spectrum Living Water

Viktor Schauberger described the attraction and repulsion of polarized forces as *the dance of creation—Nature's engine.* Polarity is the power in the universe that supports balance and creation, which includes everything from the creation of new life when male and female energies unite to the formation of a third substance when two atoms come together. Without attraction and repulsion there is no movement, and without movement there can be no life.

The Chinese described the balance between polar forces as *yin* and *yang.* The ancient yin/yang symbol provides a sense of continuous movement. The dot in the center of each indicates that neither can exist without the other. Yin energy is feminine; it is yielding, cool, dark, soft, inward, contracting, and creative. Yang energy is masculine; it is drawing, hot, light, hard, outward, expanding, and disintegrative. The ancient text known as the *Yellow Emperor's Classic of Medicine* had this to say about the balance between opposing forces:

> The principle of yin and yang is the foundation of the entire universe. It underlies everything in creation. . . . [I]t is the root and source of life and death. Yang brings about disintegration; yin gives shape to things.[1]

The ancient Chinese believed that to achieve balance, the natural ratio between these polar forces was 60% feminine to 40% masculine. Schauberger felt that the correct balance between polarities was even more weighted toward the feminine—more like two-thirds (67%) to one-third (33%). His son, Walter (a mathematician), related the proportion to *phi*, or the golden ratio, which gives the feminine 61.8%.[2] Regardless of the exact proportion, Nature's balance is tipped slightly to favor creation over disintegration.

Water is the perfect example of balance with a slightly feminine twist. In the traditional formula for water, H_2O, hydrogen, the feminine element, *yields* electrons to the bonds that hold each molecule together. Oxygen, the masculine element, *draws* electrons from each of the two hydrogen atoms to achieve electrical balance. From this perspective, the formula for an individual water molecule is two parts feminine to one part masculine.

Yet water's balance is nested in further polarity. Shared electrons spend a comparatively greater amount of time around the larger oxygen atom, causing the water molecule to assume a "V" shape with a positively charged and negatively charged side. Water itself is a polar molecule—a dipole.

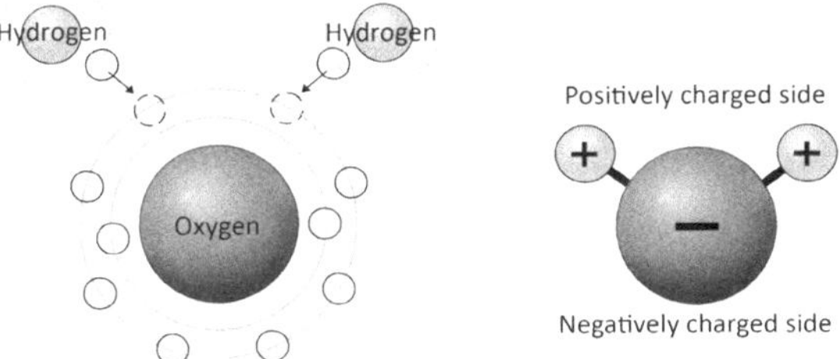

Hydrogen atoms share their electrons with oxygen atoms in the formation of water. The resulting molecule has a "V" shape with a positively charged side and a negatively charged side.

Opposing charges in the water dipole bring about another polar interaction called hydrogen bonding (see page 5). Hydrogen bonding is responsible for the interconnected network of molecules that comprise water's liquid state. Under some circumstances, the more organized liquid crystalline matrix emerges, composed of interconnected hexagonal subunits (shown in the drawing below).

Dr. Pollack and others have identified the hexagonal subunit in structured water.[3,4,5] Each subunit yields a negative charge, providing an abundance of delocalized electrons within the liquid crystalline continuum, as proposed by Szent-Györgyi (see page 8). Negatively charged electrons are balanced by positively charged protons that are forced outside areas of tightly structured water.[6]

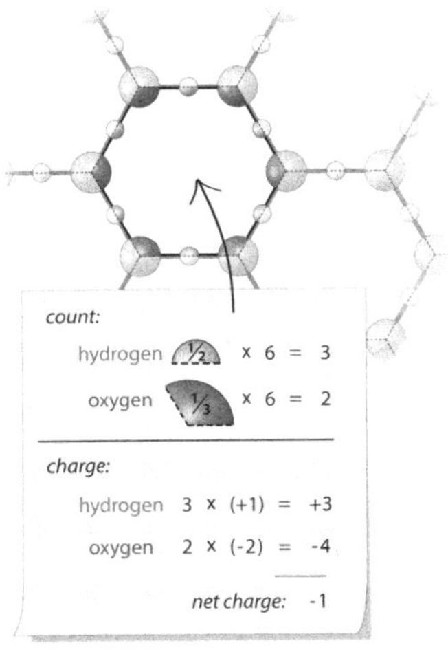

Pollack's formula for liquid crystalline water is derived by representing each hexagonal subunit as a sliceable pie. Counting the fractions of each atom inside a hexagonal unit yields the formula H_3O_2. This formula represents the feminine/masculine ratio of 60% to 40%.

Considering that atoms themselves are an intermingling of positive and negative charges, liquid crystalline water epitomizes Schauberger's *dance of creation* with a number of nested polar interactions, weighted to support creation. With the addition of positively and negatively charged ions (salts, see Chapter 4) water is energized and further balanced.

Cycles

The interplay between Nature's masculine and feminine forces results in cycles of breakdown and re-creation. Birth, the ultimate creative process, eventually results in death. Likewise, darkness finds its way to light in the continuing cycle between night and day. Masculine and feminine interaction establishes cycles in the *dance of creation* at microscopic and macroscopic levels, where all motion starts from a point of rest and returns to a point of rest.[7] The process is much like breathing.

Cycles of contraction and expansion help water to remain balanced. Flowing freely, water always finds regeneration with inwinding spiral movement, which cultivates structure and the capacity to hold energy. Outwinding spiral movement discharges energy, providing a release as water breathes. With each discharge, water assumes masculine qualities as it draws gases, minerals, and other replenishing constituents. The process keeps water healthy while nourishing all life on Earth.

Two hydrological cycles

It should come as no surprise that the Earth's hydrological cycle is actually two cycles. The masculine cycle (atmospheric water cycle) occurs above ground under the influence of the Sun, whereas the feminine counterpart (primary water cycle) occurs in the quiet, dark womb of the Earth, driven by the Earth's energy. Schauberger believed that if the atmospheric cycle was repeated without exposure to the primary cycle, water would eventually lose creative energy. In this imbalanced state, water cannot provide optimal sustenance for Earth's life forms and becomes partially responsible for their deterioration. The underground part of water's process can take hundreds, thousands, or even millions of years. The longer the journey within the Earth, the more life force water brings with it to the surface. Nature is never in a hurry.

Mature versus juvenile water

According to Schauberger, when water has completed its feminine cycle beneath the Earth, it rises to the surface on its own in the form of a true spring. He referred to this water as *mature* water, as opposed to juvenile water, which is either found on the surface or is artificially brought to the surface. True springs* result as underground cavities fill with newly generated water or as heat and pressure push water to the surface. These mechanisms are augmented by an abundance of hydrogen in the water, which makes mature water very light and easily drawn upward against the flow of gravity. Water that comes to the surface on its own typically carries a balanced blend of minerals/salts that behave differently from the minerals in juvenile water.

Minerals in mature water are in ionic (charged) form. Water organizes around these charged surfaces, building layers of liquid crystalline water. Under the right circumstances, layered regions create a coherent water array. In other words, ionic minerals serve as

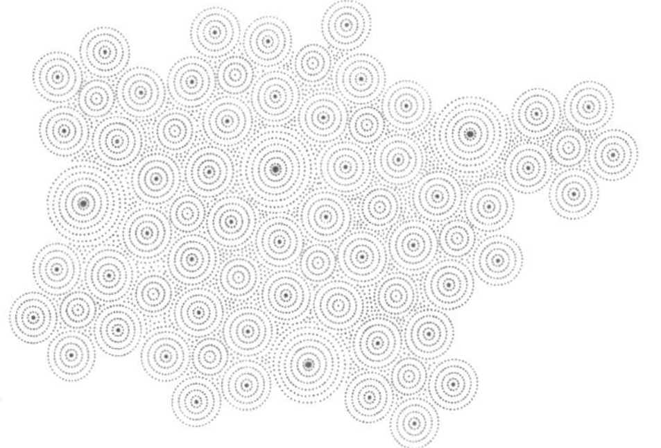

A 2-dimensional conceptualization of ionic minerals surrounded by layers of structured water, forming a more complex liquid crystalline matrix

* Seepage springs are not true springs; they result when percolating water comes in contact with a rock barrier. Water flows on top of the rock until the layer is exposed or the amount of water builds up and is pushed to the surface. Seepage springs produce juvenile water; they are dependent on precipitation.

seeds for the initiation of ordered water layers and for the development of a more complex liquid crystalline matrix. In this state, individual mineral ions are isolated from each other by layers of tightly structured water. Isolation keeps minerals from forming scale.

When juvenile water containing an abundance of carbonate minerals is brought to the surface before it has completed its underground cycle, the water is often said to be "hard." Hard water minerals (calcium and magnesium carbonate) produce deposits (scale) on faucets, showerheads, and equipment because these minerals in juvenile water tend to form large conglomerates that can appear dissolved yet they eventually gather together to form tightly cemented layers. Heat and evaporation accelerate the formation of scale. But what happens to carbonate minerals in water when the water is subsequently structured? The process of bringing structure to juvenile water either breaks up large mineral conglomerates or excludes them. As large mineral conglomerates are surrounded by layers of organized water, they drop from solution and form sediment rather than scale.

Magnetic fields have been shown under some circumstances to change carbonates from calcite to a form of calcium carbonate, called aragonite, that accumulates as acicular crystals (fragile, needlelike crystals). These tiny crystals are loosely packed. They sometimes float on the water's surface and are nothing like the dense, tightly packed calcite form of calcium carbonate that produces scale. The same changes have been noted with many methods of structuring water. Sediment on the bottom of a water container (rather than the development of scale) is a telltale sign that the water has acquired a greater degree of molecular organization or structure.

The birth of water

Beneath the surface of the Earth, water is revitalized. New science reveals that water is also *born* in the feminine, subterranean environment. Emerging evidence suggests that the Earth may be a great water generator. In fact, the capacity to create water inside the

Earth may be limitless. Researchers are discovering there is more water *inside* the Earth than previously imagined.[8,9,10] According to Japanese research, the volume of water inside the Earth may amount to five times more than is in the oceans.[11] This water is referred to as Primary water—created under intense pressure within the Earth. It was first postulated by professor Adolf Nordenskiold, who was nominated for the Nobel prize in 1901 for his work. Stephan Reiss did further research in the 1930s and successfully identified hundreds of areas of *new water* deep within the Earth's interior.[12] The Reiss method today uses mineralogy, petrology, and structural geology to locate high-pressure/low-temperature water signatures deep within the Earth. This method has produced numerous bore holes and abundant water.[13]

Universal water cycling

It is well known that the Earth regularly loses water as the Sun ionizes vapor in the upper atmosphere and releases hydrogen into space. This is evidence of another water cycle. It is likely that water cycles throughout space via a number of mechanisms. Comets are mostly ice. Meteors are up to 20% ice, and molecular clouds in space possess abundant water vapor.

Dr. Louis Frank, a professor of physics at the University of Iowa, has documented the release of enormous amounts of water vapor from comets that enter the Earth's atmosphere on a regular basis. According to his research, millions of small comets impact our atmosphere each year, showering the Earth with a continual supply of cosmic water vapor.

Full-spectrum living water

Water's birth and sojourn inside the Earth occur in the quiet, dark, feminine environment. Liquid crystalline structure develops there as it circulates within the magnetic fields of the Earth. Vortices, silicate minerals, and natural forces contribute to the development of coherence. When water is ready, it finds its way to the surface, where brief exposure to sunlight and the vibratory influences of the natural world complete the process. *Full-spectrum living water* results. Newly created and/or regenerated *full-spectrum living water* from the womb of the Earth is alive with creative force and with the patterns necessary to sustain vibrant life. Ask anyone who has had the opportunity to drink fresh spring water immediately from the Earth. Most will tell you it is an especially invigorating experience.

Water on the surface of the Earth today has been recycled so many times without energetic revitalization that it is difficult for it to carry life-sustaining energy. But there are many simple things you can do to enliven and mature the water you drink and the water you use in your home and garden. Part II discusses a variety of ways to create *full-spectrum living water.*

Image Credits (Chapter 2)

Page 19 Water hexamer as a sliceable pie:
Pollack, G. (2013) *The Fourth Phase of Water,* Ebner & Sons

References (Chapter 2)

1 As quoted in: *Chinese Civilization: A Sourcebook*, 2nd ed., (1993). P. Ebrey, Free Press, pp. 77-79.

2 Bartholomew, A. (2005). *Hidden Nature: The Startling Insights of Viktor Schauberger*. Adventures Unlimited Press, p. 53.

3 Pollack, G. (2013). *The Fourth Phase of Water: Beyond Solid, Liquid, Vapor*, Ebner and Sons, pp. 59-60.

4 Lippincott, E. Stromberg, R., Grant, W. and Cessac, G. (1969). Polywater. *Science*, 164, pp. 1482-1487.

5 Jhon, M-S. (2004). *The Water Puzzle and the Hexagonal Key*; Uplifting Press.

6 Pollack, G. (2013). *The Fourth Phase of Water: Beyond Solid, Liquid, Vapor*, Ebner and Sons, pp. 52-54 and 203-210.

7 Russell, W. (1953) *A New Concept of the Universe*. The Walter Russell Foundation.

8 Pearson, D.G. et al. (2014). Hydrous Mantle Transition Zone Indicated by Ringwoodite Included Within Diamond. *Nature*, 507(7491), pp. 221-224. Available online:

http://www.nature.com/nature/journal/v507/n7491/full/nature13080.html

9 Schmandt, B. Jacobsen, S. Becker, T. Liu, Z. and Dueke, K (2014). Dehydration Melting at the Top of the Lower Mantle. *Science,* 344(6189), pp. 1265-1268.

10 Murakami, M., Hirose, K., Yurimoto, H., Nakashima, S. & Takafuji, N. (2002).Water in Earth's Lower Mantle. *Science* 295, 1885–1887. Abstract online: http://www.sciencemag.org/content/295/5561/1885.abstract

11 Murakami, M., Hirose, K., Yurimoto, H., Nakashima, S. & Takafuji, N. (2002). Water in Earth's Lower Mantle. *Science* 295, 1885–1887. Abstract online: http://www.sciencemag.org/content/295/5561/1885.abstract

12 King, M. (2014). Primary Water Research: Demonstration Sites and Laboratory Research. Available online: http://www.primarywaterinstitute.org/images/pdfs/M_King_Primary%20Water%20Research.pdf

13 King, M. (2014). Primary Water Research: Demonstration Sites and Laboratory Research. Available online: http://www.primarywaterinstitute.org/images/pdfs/M_King_Primary%20Water%20Research.pdf

Chapter 3

Vortices and Spirals

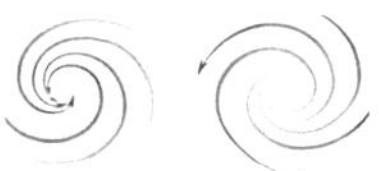

Spiral motion is everywhere—not just in water but also in the air currents that shape our weather, in the growth of plants, in atoms and galaxies, and in the spiral of our DNA. Vortices are a perfect example of Nature's use of polarity to achieve balance. They are also Nature's finest organizing forces, and they are essential in the creation of *full-spectrum living water.*

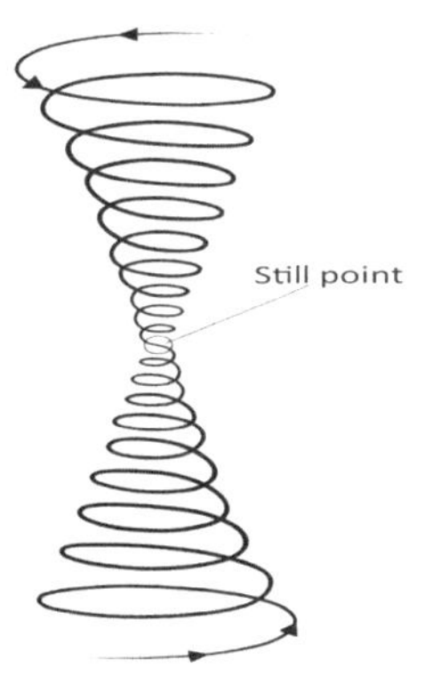

Twin opposing vortices with still point in the center and reversal of spin

According to Walter Russell (1871–1963), who presented a unified theory of energy dynamics known as Russellian Science, *all of nature is composed of twin opposing vortices.* For every vortex drawn downward or outward, another vortex exists in the opposite direction.

An interesting illustration of the power of opposing vortices is recounted by Viktor Schauberger in an article he wrote for *Implosion Magazine*:

> It was raining in sheets. I lay on the float, looking down into the powerful vortex. As though falling from a gutter, the rainwater flowed from the upturned brim of my hat and into the hole. According to the conventional theory of gravity it ought to have fallen into the vortex. However, it did not. As it fell it widened out into a cone, broader at the base, and in this way a hat-shaped funnel formed above the lower vortex. . . . I observed this extraordinary phenomenon very intently. But not for long because suddenly an ice-cold jet of water struck me right in the face.

Opposing vortices create opposing forces, and it is the levitational force that salmon use to propel them high into the air as they bound up waterfalls during annual spawning runs. Aligning themselves within the unseen vortex, salmon are capable of jumping beyond their physical limits. Trout use this same force (on a horizontal rather than a vertical axis) to remain stationary in a strong current. The shape of the trout, as well as the passage of water through its gills, is reported to create vortices in front of and behind it. Trout are literally sucked forward despite a strong current in the opposite direction. The potential for levitation/suction increases with the water's velocity. It is also strengthened by the presence of rapidly spinning elements in the water (see ormus, page 34).

Levitational/suctional forces are equal to, if not more powerful than, gravitational forces. Consider how buildings and other large objects are uprooted by the suction of a tornado or hurricane. Schauberger and others used these forces to make levitating devices that, admittedly, mainstream science has never understood. His technology and most of his records were destroyed after World War II. Devices based on similar technology by later inventors have been suppressed.

Vortices gather and discharge energy. The gathering, creative force imposes molecular organization as it condenses matter (revealed by Schauberger and Pöpel in their experiments where the water appeared to leave the walls of the pipe [see Chapter 1]). This creates suction/implosion. Vortices can also discharge energy, depending on the type of motion involved.

Explosion versus implosion

Two basic types of motion are found in the natural world that both Schauberger and Russell described. One is the outwardly expanding spiral motion that begins slowly at the center and increases in speed toward the periphery. This type of motion is referred to as *centrifugal* motion. It is the same principle used in the spin cycle of a washing machine where the contents of the machine are thrown to the periphery as it spins faster and faster to expel water. This type of motion produces heat and friction as it indiscriminately flings energy outward. It is *explosive* and deconstructive (masculine). Nature uses centrifugal movement for breakdown and decomposition. Unfortunately, it is also used in most of our current energy-generating technologies—from combustion engines to nuclear reactors. Much of the energy that is produced is wasted.

The other type of spiral motion is incorporated during Nature's creative processes. It is the *inwinding*, *implosive* (feminine) energy that begins from the outside and winds inward, gathering, condensing, cooling, and organizing as the velocity increases. This type of motion is referred to as *centripetal* motion. It is a constructive, friction-reducing form of movement that *generates* energy. According to Schauberger and Russell, if we were to use this type of inwinding motion in our energy-generating devices, we would have all the energy we need, with little if any ongoing cost. Both kinds of motion can be used to generate energy; both employ spiraling movement. The difference lies in whether the vortex winds inwardly (producing suction) or outwardly (requiring pressure).

Inwinding centripetal motion
Characterized by:
suction
absence of friction
temperature drop
growth and development

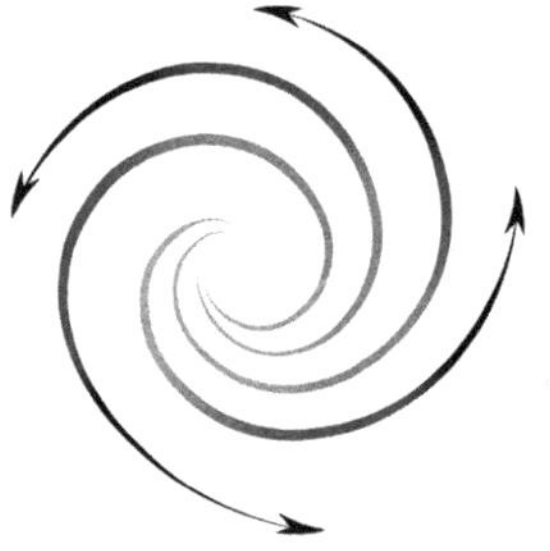

Outwinding centrifugal motion
Characterized by:
pressure
friction
temperature rise
deterioration

Nature uses vortices to minimize energy expenditure. Like trout, insects create vortices in front of and above them to support flight. Their wings, which appear to work so hard, require minimal effort within the suction of a vortex. The curvature of a beetle's wing is a perfect example of capturing lifting energy. Spiral movement also plays a role in how sap is drawn upward in the tallest trees—a feat that conventional science still cannot explain.[1]

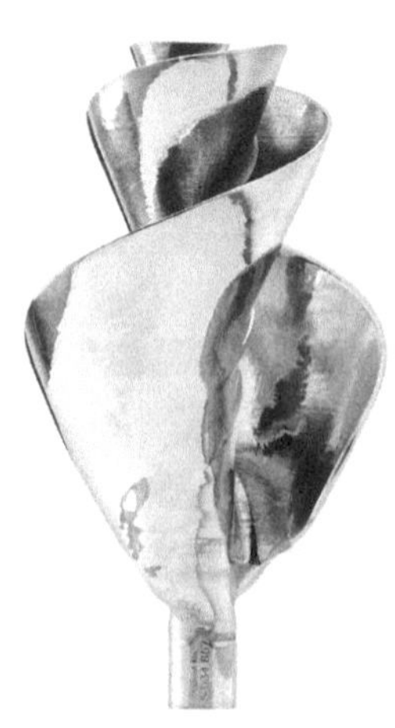

Lily Impeller

When water is *guided* (rather than forced) using flow forms that encourage centripetal motion, we can bring organization and molecular structure to the water with little effort. A variety of devices inspired by Nature are being crafted with this idea in mind. One example is the Lily Impeller designed by Jay Harman. It draws water inward rather than pushes water outward as with a propeller. This gathers energy and organizes water's molecular array. According to Harman's research, a tiny 6 inch impeller is able to circulate thousands of gallons of water and reduce the need for sanitation chemicals during storage. Logarithmic spiral-shaped rotary devices can reduce energy requirements up to 80% and noise up to 75% over conventional rotors. These types of devices have a wide range of applications, from fans to water turbines.[2]

The Ranque-Hilsch effect

Cooler air condensing in the center of a spiral pipe is known as the Ranque-Hilsch effect. Schauberger demonstrated a similar effect in water by showing that water was cooler after it had passed over a rock due to the vortex created as water spins past obstacles. The temperature at the center of a vortex is cooler than at its periphery. This is one reason moving water stays cool despite warmer ambient temperatures. Centripetal movement keeps water cool, energized, and organized.

Characteristics of a vortex

Johannes Kepler (1571–1630), a German astronomer and mathematician, discovered that the inner layers of a vortex move more rapidly than the outer layers, similar to the way planets circle the sun—faster when they are closer. In water, layers glide past each other with clearly defined boundaries.[3] Each vortex also has a peripheral margin making it a separate energetic entity within its surrounding environment, much like an organ within an organism. Vortices in water act as antennas and sense organs opening and closing for the exchange of information from the universe. And because the Earth's surface is 70 to 80% water, its capacity to receive and store information is amplified. This is one of the reasons life has flourished and why it continues to evolve on Earth.

Close observation reveals that vortices also have a natural pulsation—a rhythm of contraction and expansion. As they contract, gathering energy, they reach downward. The expansion that follows literally draws water and its contents upward and outward. This is how mineral-rich sediment from the bottom of the ocean is brought to the surface and recycled. Vortices describe the yin and yang of the universe. They are the continual in breath (gathering) followed by the out breath (release) that governs all life.

Vortices and the law of thermodynamics

The second law of thermodynamics states that a system continues to degenerate if no outside energy is infused. This is called entropy—the progression toward disorder represented by a gradual decay and loss of organization. Vortices appear to defy the laws of thermodynamics. Schauberger demonstrated that under some circumstances vortices *generate* energy. He built a water-powered generator he called a "centripulser." It used both centrifugal and centripetal forces. Water was spread out from a central hub through spiral pipes that were shaped like kudu antelope horns, creating double spirals. In Schauberger's centripulser, the force of the water leaving the generator was so powerful that it penetrated both concrete and steel. One observer recalls:

> The machine produced an almost inaudible sound and then a "pftt" in the same instant and the water pierced right through a 4 cm. thick concrete slab and a 4 mm. thick super-hardened steel plate with such force that the water particles, invisible to the eye due to their high velocity, penetrated right through all clothing and were experienced as lightning needle-pricks on the skin.[4]

Torsion fields: paired vortices

Beginning in the 1950s, Russian scientists* identified torsion fields (paired vortices forming a torus) as conduits for the flow of energy and information in space.[5] They determined that stars, planets, and

* The work of Dr. Nikolai Kozyrev (1908–1983), a renowned Russian physicist, stimulated hundreds of papers in Russia on the science of torsion fields. He and others have defined the nature of torsion fields.

galaxies exchange energy and information at super-luminal speed (faster than the speed of light) through torsion fields. According to this research, torsion fields are spiraling channels that serve as energy/information guides.[6] Not only do they facilitate the flow of information; they may also be the means by which the unlimited energy of the universe is translated into 3-D form (light, heat, electromagnetism, bioenergy/life force, etc.).[7]

The torsion field: paired vortices forming a torus

Torsion fields are caused by any form of movement or rotation. A 2013 paper by Nassim Haramein offers mathematical evidence that protons are black holes,[8] described as *singularities*, at the center of a torsion field. Haramein's work reveals what many others have postulated—that protons are not really particles at all; rather they are tiny torsion fields. His work explains why electrons do not crash into the center of atoms (something modern physics has been unable to explain). Subatomic torsion fields continuously replenish their energy from the universal energy field.

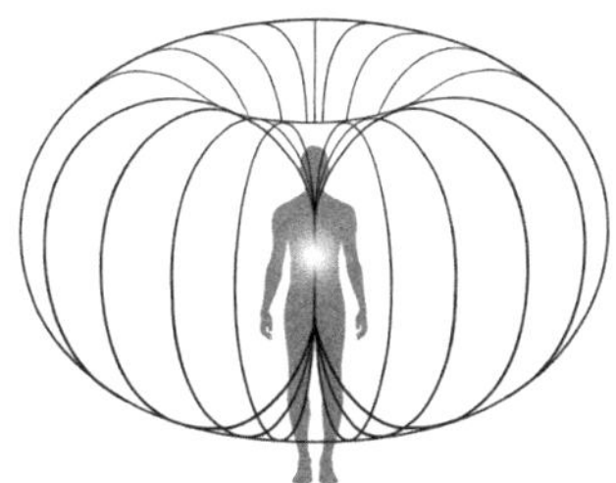

Biological organisms are also surrounded by torsion fields with the potential to continuousy replenish their life force from the universal source field.

Torsion waves (emanations from torsion fields) are the equivalent of scalar waves, as described by Nikola Tesla in the early 1900s. Scalar technology and torsion physics are widely researched in some parts of the world, though still not acknowledged by traditional Western science. This cutting-edge research reveals that torsion/scalar energy is present everywhere, keeping the entire cosmos pulsating and alive. Harnessing this energy has been the driving force behind the free energy movement for years. The key appears to be in torsion physics and vortex math. Vortices have long been considered gateways between levels of energy, which helps to explain another vortex phenomenon with significance for water known as ormus.

Ormus

The term "ormus" evolved from the original acronym for *orbitally rearranged monoatomic element* (ORME), which was used by David Hudson to describe a unique monoatomic state in certain elements.[9] The original ormus elements are a group of metallic elements in the center of the periodic table: ruthenium, rhodium, palladium, silver, osmium, iridium, platinum, and gold. Other elements besides the original eight metals are now thought to exhibit this state.

In the ormus state, an element's outer electrons pair up with themselves rather than seeking to pair with other atoms. This arrangement, called Cooper pairing, causes each atom to spin rapidly—reminiscent of an ice skater as her arms are drawn inward. Theoretically, this reduces an atom's size and weight. When atoms enter the high-spin/ormus state, they are no longer able to form chemical bonds with other atoms. However, large groups of them may gather to form *coherent domains* with superconductive properties and little resistance to the transmission of energy and information. This is possible because these elements in their ormus state develop an antimagnetic field called a *Meissner field.*

Unlike electrically conductive substances, the energy in a superconductive domain is transferred from Meissner field to

Meissner field,[10] which explains how the presence of ormus in water may contribute to its structure and enhanced ability to conduct energy and information. Meissner fields surrounding ormus elements actually repel water molecules, but in doing so, water molecules are required to become more organized. Water with an abundance of ormus has superconductive potential. Within the human body it is capable of transmitting signals and other information with close to zero resistance.[11] This may explain why some who consume ormus on a regular basis experience new levels of well-being and awareness.

Elements in their ormus state have another unique characteristic—they levitate. Their spin and resulting Meissner field create a levitating force against the Earth's magnetic field. Ormus elements lose 4/9ths (44%) of their weight,[12] which contributes to the levitational forces in *mature* water that rises to the Earth's surface against the flow of gravity. Elements in the ormus state are plentiful in natural springs. According to Barry Carter, an expert on ormus, "[O]rmus elements are twice as abundant at the spring as they are 300 meters downstream."[13] This suggests how plentiful they are in mature water, but it also reveals how rapidly they dissipate as water is exposed to light and heat above ground. The addition of salts helps to hold ormus elements in the water.[14]

Although many methods have been devised to create the ormus state,[15] spinning water within a magnetic field (referred to as a magnetic trap) is one of the easiest.[16] The combination of vortex and magnetic field induces the high-spin state in elements prone to Cooper pairing. By incorporating salts with a high content of these elements, the potential for ormus in the resulting water increases and water's structure becomes more stable.

Revitalizing water using vortices

The new sciences have opened the way for a broader understanding of the role of vortices in the channeling and translation of energy. From the spinning of each subatomic particle to the spiraling of galaxies, vortices gather and cohere energy. They are the self-organizing engines of the universe and are extremely important when it comes to water. When water is allowed to flow in ways that encourage freedom of motion, it naturally spins and creates interacting vortices. The process produces a greater degree of structure and gathers creative energy. In Part II, you will learn a variety of methods for using vortices to structure water and methods that enhance water with ormus.

Image Credits (Chapter 3)

Page 30 Lily Impeller, courtesy of Pax Scientific; www.paxscientific.com

References (Chapter 3)

1 Coats, C. (2001). *Living Energies: Viktor Scahuberger's Brilliant Work with Natural Energy Explained* 3rd revised edition, Gill & Macmillan Ltd Dublin Ireland, pp. 246-247.

2 Harman, J. (2013). *The Shark's Paintbrush: Biomimicry and How Nature Is Inspiring Innovation*, White Cloud Press.

3 Schwenk, T. (1996). *Sensitive Chaos* revised second edition, Rudolf Steiner Press, Plate 37.

4 Bartholomew, A. (2005). *Hidden Nature: The Startling Insights of Viktor Schauberger* Adventures Unlimited Press, p. 254.

5 Akimov, A. and Shipov, G. (1996). Torsion fields and their experimental manifestations. Proceedings of International conference: New Ideas in Natural Science.

6 Akimov, A. and Tarasenko, V. (1992). Models of polarized states of the physical vacuum and torsion fields. *Soviet Physics. Journal 35:3.* p. 214.

7 Wilcock, D. (2012). *Divine Cosmos.* Sections 1.4 and 1.5. Online book available at www. http://divinecosmos.com

8 Haramein, N. (2013). Quantum Gravity and the Holographic Mass. *Physical Review & Research International* 3(4): 270-292. Available online: http://hiup.org/wp-content/uploads/2013/01/1367405491-Haramein342013PRRI3363.pdf

9 Hudson, D. and Concord Research Corporation, UK Patent Application GB2 219 995A, Filed 6-21-89. Available online: http://www.subtleenergies.com/ormus/patents/ukpatent.htm

10 Quantiki Website. Superconductivity http://www.quantiki.org/wiki/index.php/Superconductivity

11 Personal communication with John Milewski—Internationally recognized leader and consultant in advanced Materials Science.

12 Milewski, J. (2003). ORMUS is a gas. Available online: http://www.hbci.com/~wenonah/hudson/ormusgas.htm.

13 Carter, B. Living Water, Vital Air. Available online: http://www.subtleenergies.com/ORMUS/tw/livwater.htm

14 Carter, B. Living Water, Vital Air. Available online: http://www.subtleenergies.com/ORMUS/tw/livwater.htm

15 Hudson, D. and Concord Research Corporation, UK Patent Application GB2 219 995A, Filed 6-21-89. Available online: http://www.subtleenergies.com/ormus/patents/ukpatent.htm

16 Carter, B. Living Water, Vital Air. Available online: http://www.subtleenergies.com/ORMUS/tw/livwater.htm

Chapter 4
The Ocean and Salt

René Quinton

In the late 1800s, a French scientist, René Quinton (1866–1925), became interested in the similarity between the oceanic environment and the saline environment within our bodies. His research uncovered a striking similarity between the composition of the ocean and the composition of our internal fluids. Quinton's years of investigation demonstrated that an organism's health depends on maintaining its own *internal ocean.*

Consider the places where a dilute ocean is preserved in the human body. Blood is a dilute mirror of the ocean.[1] Amniotic fluid, lymph fluid, and extracellular fluids are also close replicas. The slightest deviation from optimal parameters in these fluids creates imbalance and ill health.

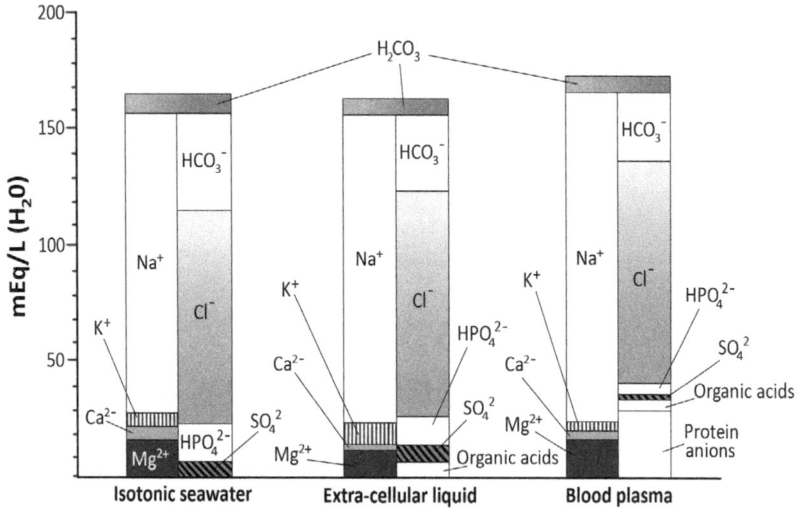

Comparison of electrolyte compositions in dilute ocean water, blood plasma, and extracellular fluid (Adapted from: *Significance of the Body Fluids in Clinical Medicine,* 2nd ed. 1955, by Leaf and Newburgh).

Quinton published his early research in a book titled *L'eau de mer milieu organique* (*Seawater Organic Medium*), establishing the relationship between seawater and blood plasma. He demonstrated that human blood and dilute seawater are interchangeable. On one occasion, Quinton removed the entire volume of blood from a dog and replaced it with filtered, dilute seawater. After a short recovery, the dog exhibited more youthful energy than prior to the transfusion.[2] This experiment was duplicated in 1969 by the Delalande Medical Research Center in France.[3]

Quinton developed a method for harvesting, diluting, and cold sterilizing seawater to preserve its living qualities. In 1907, this solution was made available to doctors as an oral and injectable treatment. At that time, a daily newspaper published the following statement (translated):

> Pasteur's work provided us with a theory of disease. That of Quinton has provided us with a theory of health. . . . What is Pasteur's serum? It is a serum for and against a

> particular disease, a serum that attacks a given microbe and none other. What is seawater? It is a serum that attacks no microbe in particular; instead it gives the organic cell the strength to fight against all of them.[4]

Numerous free clinics were opened in France and eventually in many parts of Europe and Northern Africa where injections of *Quinton's Plasma* were administered according to his protocols. Records show that this serum saved the lives of tens of thousands of individuals with a success rate well over 90% for a variety of conditions, including cholera, psoriasis, enteritis, dyspepsia, failure to thrive, and malabsorption.

Blueprint for life

The ocean is the amniotic fluid of the Earth and a storehouse of genetic information. As Quinton observed, seawater can literally *ignite* and *renew* genetic potential. Perhaps this is why the aging process does not occur in the sea in the same way it does on land. A comparison of the cells of an adult whale and the cells of a newly born whale show no evidence of the changes that are obvious between adult and newborn land mammals.[5] Until recently, life in the ocean has been exempt from the maladies experienced by life outside its protection. Chronic diseases and cancers are rarely found in the ocean as they are on land.

Quinton demonstrated that the ocean may be the key to restoring compromised genetic codes in biological organisms. Case studies conducted in Quinton's clinics revealed that seawater provided recipients with the capacity to overcome many genetic imperfections. When supplied during their pregnancies, mothers who had delivered up to eight previous stillborn or genetically abnormal children *all* delivered healthy, normal infants. Thousands of children who received Quinton's filtered seawater recovered from genetic maladies.[6]

Seawater is a blend of organically complexed minerals, enzymes, vitamins, proteins, DNA, RNA, organic acids, and other undefined biochemical constituents. It contains the full complement of minerals and trace elements, yet its life-supporting qualities are not found in the nutrients alone. *Information* (patterns and genetic codes) held in the water is equally important. Quinton discovered that heating or drying destroyed the efficacy of the solution.[7] When the water was removed, the *information* it contained was lost.

Extracellular fluid: our internal terrain

One of the most important fluids within the human body is the fluid that bathes each cell, known as *extracellular fluid*. Every cell is connected through this fluid that transports nutrients and delivers signals.[8] It is considered the *terrain* for the cells of the body and is a near-perfect match for isotonic* seawater. The term *terrain* is borrowed from agriculture, referring to the soil. It is widely understood that soil is more than minerals and organic matter providing sustenance and physical support for plant roots. Healthy soil is teaming with microbes, organic acids, enzymes, and countless components that science has not yet fully characterized. Together, these substances possess a living quality. The same is true of the ocean and our internal fluids.

Extracellular fluid is the *terrain* surrounding every cell. When this fluid is optimally maintained, cells function for a long time. This was shown by Nobel Prize laureate Alexis Carrel, who demonstrated that cells can live almost indefinitely when they are kept in an optimum terrain. Dr. Carrel kept embryonic chicken heart tissue alive for over 30 years using chicken serum and a special saline solution that was changed every day. Many believe the saline portion of the solution, which Dr. Carrel never revealed, was Quinton's filtered seawater. Dr. Carrel was a longtime acquaintance of the director of Quinton's largest clinic.[9] Carrel's work was never duplicated with man-made saline solutions, which lack the *living* quality of seawater.

---

* The term *isotonic* refers to a dilution that matches the saline concentration of human fluids. The human body has a .9% saline concentration, at which cells neither swell nor shrink. Undiluted ocean water has a saline concentration of 3.3%.

Quinton's patients were repeatedly able to establish homeostasis based on restoration of their internal terrain. The German physician Röpffer investigated a variety of changes in internal terrain due to the intake of seawater. In particular, he noted changes in the body's pH balance:

> [F]or hyperacid organisms, a dramatic decrease in acid values has been recorded. No case has remained the same. It can therefore be stated that in case of a systemic inconstant acidity, a cure of drinkable ocean water causes the normal acidity to be recovered. In particular for gastritis due to dietary errors or alcohol and nicotine abuse, remarkable results have been obtained.[10]

Ocean/marine plasma today

Quinton's Marine Plasma has been available for medicinal and nutritional use since the early 1900s. Doctors in Europe used it for a long list of conditions until the 1970s, when it was removed from the French Medical Dictionary because it was not heat sterilized. The French army used it as a blood plasma replacement for many years.** Although Quinton is largely unknown, his legacy is well remembered in parts of Europe. Today, several companies harvest and package seawater in a manner similar to Quinton's original protocol to preserve the living qualities of the solution. Many health-oriented individuals still collect their own seawater to use in cooking and for regular consumption.

Oceans: salt of the Earth

The ocean (like the fluids in the human body) is a dynamic, living crystal, continuously exchanging energy and information with its surroundings. One of the reasons water in the ocean and water in the human body are both structured is because they contain salts (minerals). Salts are an integral part of the healthy water matrix;

** A list of the pre-1975 accepted medicinal uses for Quinton's Marine Plasma can be found in the article on the *Dancing with Water* website: www.dancingwithwater.com.

they help to create and hold liquid crystalline structure. However, the structure that results in water with salts is different from the structure identified by Dr. Pollack as discussed in Chapter 1. The unique fourth phase of water he identified excludes *everything.* That's why he called it EZ (exclusion zone) water. So how can water with salts still be a liquid crystal?

Water forms hydration layers (which Pollack referred to as exclusion zones) around charged surfaces.[11] These highly ordered layers of water have a negative charge; they are surrounded by proton-rich water with a positive charge that attracts nearby negatively charged layers. As salts ionize and disperse, a unique crystalline array held together by layers of structured water emerges. Depending on the minerals, any number of crystalline patterns may result. Thus, the addition of seawater and/or unprocessed salt provides a framework for the development of a more complex and coherent liquid crystalline matrix.

"Salts"

In the absence of the ocean, unprocessed salt provides a balanced blend of minerals while introducing influences from the sea essential to our well-being. Water and salts are obvious companions. Neither is complete without the other. Salts are incapable of delivering their energy without water. Similarly, water without salts does not conduct energy. The mineral-rich saline environment provides an optimum medium for the transfer of energy and information. You could say that *minerals anchor the life force* in water. There are many salts (combinations of acidic and alkaline ions held together by electrostatic bonds) besides sodium chloride. The vast number of combinations referred to as "salts" include seven major ions that make up over 99% of the salts found in seawater.

The seven major ions in seawater with their percentages of the total

From Marine Biology (10th edition), by Peter Castro and Michael Huber

chloride (Cl^-)	55.03%
sodium (Na^+)	30.59%
sulfate (SO_4^{2-})	7.68%
magnesium (Mg^{2+})	3.68%
calcium (Ca^{2+})	1.18%
potassium (K^+)	1.11%
bicarbonate (HCO_3^-)	0.41%

Refined (table) salt

Unlike natural, unprocessed salt, refined salt (NaCl) is imbalanced and aggressive. Its aggressive nature is due to the fact that it has been stripped of many components. In nature, minerals such as calcium, magnesium, and potassium balance the sodium. Sulfates and bicarbonates balance the chloride. During the refining process, these components are removed as *impurities*, leaving a skeleton of the original salt.

Refined salt has given "salt" a bad name. The human body responds to refined salt by drawing water from inside its cells to dilute excessive, unbalanced sodium that accumulates in extracellular fluid. The reaction causes water retention (water held in the extracellular fluid outside cells). Refined salt contributes to high blood pressure, diabetes, and a whole list of maladies. No wonder doctors often recommend a salt-free diet. But it is not a salt-free diet that is necessary; what is necessary is natural, unprocessed salt.

Low-salt diets have adverse effects on numerous metabolic markers, including insulin resistance. These diets have been associated with heart disease and mineral deficiencies.[12] The truth is, many individuals suffer from a *lack of salt*, even though they consume copious amounts of refined (table) salt. The deficit of balanced, unprocessed salt in the diet causes salt cravings that can never be satisfied with refined salt.

The quality and variety of ocean minerals in natural proportion are more valuable than any refined or manufactured compound. Sports drinks and man-made electrolyte solutions supply a few minerals without the full complement found in seawater and/ or unprocessed salt. Dissolving a bit of unprocessed salt in water is far superior to expensive, man-made solutions intended to replenish electrolytes following exercise, sickness, or detoxification.

Bicarbonate salts

The ocean is a buffered solution with the ability to maintain a stable pH between 7.5 and 8.5. The acid-base balance in the ocean is maintained the same way it is maintained in the human body—via a system referred to as the *bicarbonate buffering system.* In the ocean, the bicarbonate buffering system neutralizes an enormous amount of acidic waste produced by plant and animal life on a daily basis. In the human body, the same buffering system keeps the pH of the blood within a tight range (7.35–7.45) and neutralizes acidic metabolic wastes. This is another illustration of how our body's fluids are a mirror of the ocean.

Neutralization of acids depends on the presence of buffers. They stabilize pH. For example, if you add a weak acid to a glass of water that has no buffering capacity, the pH will immediately drop. But if you add the same amount of acid to a glass of buffered water, the pH will barely change. Buffers also stabilize water's structure as discovered by Dr. Pollack:

> EZs [exclusion zones: the equivalent of organized water] are especially robust in the presence of buffers, and observed (albeit smaller) in the presence of physiological salt levels. . . . We found indeed that buffers exerted a stabilizing influence on the EZ, and also that in the presence of buffers, EZs could be seen adjacent to both negatively charged and positively charged surfaces.[13]

In the ocean (and in the water of the human body) buffers are predominantly bicarbonates. The bicarbonate buffering system (like all buffers) is composed of a weak acid (carbonic acid: H_2CO_3) and a weak base (bicarbonate: HCO_3^-). When acids are added to the solution, the bicarbonate ions reassociate with H^+ ions and the pH comes back to the specified range. The bicarbonate buffering system depends on the availability of carbon dioxide in the water (see Chapter 5).

$$CO_2 + H_2O \gg H_2CO_3 + H_2O \gg H^+ + HCO_3^-$$

Carbon dioxide + water » Carbonic acid + Water » Hydrogen ion + Bicarbonate

The greatest limiting factor for neutralizing acidic waste in the body is the availability of carbon dioxide and the resulting bicarbonates. Without them, aging and disease progress at accelerated rates. Diet, pollution, exercise, and age all draw on bicarbonate reserves and limit the body's ability to neutralize acids. Adding bicarbonates to your drinking water can balance pH (help neutralize metabolic acids) and stabilize water's structure.

Salt, negative ions, and cleansing

Have you ever noticed that granulated salt (without anti-caking agents) tends to form clumps in a warm, humid environment? This is because water and salt are attracted to each other. Warm temperatures cause tiny atmospheric water droplets to be attracted to salt. When the water evaporates, salt returns to its crystalline state. At the ocean, this process takes place over and over again as water breaks salt's ionic bonds and liberates a field of ions that energize plant and animal life while neutralizing pollutants—another of Nature's miracles.

The interaction between water and salt produces ions—one of the simplest ways for Nature to neutralize pollutants. Even though the oceans have received an onslaught of contamination, their high salt concentration provides a way for Mother Nature to respond. It is one reason sea salt continues to be a safe option for those who choose not to use processed table salt. The production of natural salt requires months of drying in ponds, where it is concentrated as it is exposed to the Sun. Under these conditions, countless negative ions are released. They neutralize innumerable contaminants (including radiation) in much the same way antioxidants neutralize free radicals in the human body.

The ocean: our home

Many people describe a visit to the ocean like *going home*. While at the beach, they walk barefoot on the Earth; they breathe ion-rich air and absorb nutrients from the ocean through their skin as they swim. And who would have guessed that the occasional unintended gulp of seawater was also in their best interest.

In Part II we will discuss how to use microfiltered seawater, unprocessed salts, and bicarbonates to structure and augment *full-spectrum living water*.

Image Credits (Chapter 4)

Page 39 René Quinton; Quinton Family Archives

References (Chapter 4)

1 Dittman, R. (2006). Bio-Terrain, Evolutionary Biology and the Practice of Medicine in the Early 1900s: An Intro to René Quinton's Marine Plasma. *Explore!* 15(4).

2 Quinton, R. (1912). *L'eau de mer milieu organique* . Reprinted: Ed. ENCRE 1995. pp. 169-170.

3 Delalande Medical Research Center (France) under the directorship of Dr. B. Pourrias and Dr. G. Raynod – May 1969. Available online. Experiment on a dog with ocean (Quinton) plasma during a stage of hemorrhagic shock

4 Quoted in: *Quinton, La cure d'eau de mer* by Jean-Claude Secondé. Editions Chariot d'Or, 2011.

5 Murray, M. (1976). *Sea Energy Agriculture*, Printer's Ink.

6 Mahé, A. (1962). *Le Secret de Nos Origines* - 2nd edition, Le courrier du Livre.

7 Dittman, R. and Brugioni, R. (2006). Evolutionary Development of Our Internal Ocean: Restoring Bio-Terrain with Quinton Marine Plasma. *Explore!* 15(6). Available online: http://www.equilibrauk.com/Quinton/Dittman2.pdf.

8 Pischinger, A. (2007). *The Extracellular Matrix and Ground Regulation: Basis for a Holistic Biological Medicine.* North Atlantic Books, pp. 3-11.

9 Carrell, A. Nobel lecture. Available online: http://nobelprize.org/nobel_prizes/medicine/laureates/1912/carrel-lecture.html

10 Röpffer, as quoted in: Comparative Study of the Therapeutic Properties of Seawater Preparations by André Passebecq, and Jean-Marc Soulier, Available online: http://oceanplasma.org/documents/passbecsoulier-e.html

11 Chai B, Zheng J, Zhao Q, and Pollack G. (2008). Spectroscopic Studies of Solutes in Aqueous Solution, *Journal of Physical Chemistry A* 112(1) pp. 2242-2247.

12 Brownstein, D. (2012). *Salt Your Way to Health,* 2nd ed. Medical Alternative Press, p. 53.

13 Zheng, J. Wexler, A. and Pollack, G. (2009). Effect of Buffers on Aqueous Solute-Exclusion Zones Around Ion-Exchange Resins. *Journal of Colloid Interface Science,* 332(2): pp. 511–514. Available online: http://www.ncbi.nlm.nih.gov/pmc/articles/PMC2677409/

Chapter 5

Hydrogen, Oxygen, and Carbon Dioxide
The Gases Dissolved in Water

Water is made of hydrogen and oxygen. Contact with the atmosphere means these gases are also *dissolved* in water. Carbon dioxide is present in water too. Each gas has a unique role. Depending on circumstances, the amount of these gases fluctuates to maintain a natural healthy balance. Enrichment with hydrogen, oxygen, and carbon dioxide can create water with therapeutic properties.

Albert Szent-Györgyi

Hydrogen

Dr. Albert Szent-Györgyi's Nobel Prize–winning work was based on identifying the reactions that liberate energy from hydrogen. He explained one of the basic principles of biology: hydrogen and oxygen interact in a delicate balance releasing energy and delivering it to cells in tiny portions. Dr. Szent-Györgyi said: "Hydrogen is, in fact, the only fuel the body knows. The foodstuff, carbohydrate, is essentially a packet of hydrogen, a hydrogen supplier and hydrogen donor, and the main event during its combustion is the splitting off of hydrogen. So the combustion of hydrogen is the real energy-supplying reaction."[1]

Hydrogen is a form of energy currency. It is the fuel that provides biological energy in a variety of ways:

- In ionic form, hydrogen is a source of protons (H^+) and electrons (H^-). Both are necessary in the mitochondria for the production of ATP—the body's primary source of energy.
- Molecular hydrogen (H_2) can support ATP production too, independent of the mitochondrial electron transport chain.[2] A reaction called the Jagendorf reaction bypasses the proton gradient by establishing a hydrogen-rich environment.
- Hydration (the movement of hydrogen into cells) generates hydroelectricity.[3]

Since the turn of the century, hydrogen has taken its place in modern medicine.[4,5] As the smallest and most mobile element, hydrogen moves quickly into cells and fluids.[6] It can be found in the blood within minutes after consumption in water.[7] Hundreds of articles have been published on its therapeutic value, including antioxidant and anti-inflammatory effects,[8,9] protection against radiation damage,[10,11] stimulation of energy metabolism,[12] enhanced athletic performance,[13] improved glucose metabolism,[14] protection against brain trauma[15] and cognitive disorders, and remediation of many age-related symptoms.[16]

Hydrogen pools

Dr. Szent-Györgyi discovered that the organs of the body store significant amounts of hydrogen. He coined the term *hydrogen pooling* and revealed that the organ with the greatest capacity to store hydrogen is the liver. The liver utilizes hydrogen to quench the free radicals produced by detoxification. During the process of aging, the pools of hydrogen in our organs become depleted.

When hydrogen pools are depleted, the body draws hydrogen from its most plentiful source—water inside our cells.[17] In this way, cellular water is often sacrificed to provide necessary hydrogen. It is one reason aging is accompanied by a loss of water. However, when hydrogen is made available, hydrogen pools can be replenished with a resulting reduction in free radical stress. Cognitive functions may improve, pH balance is often restored, and as you might expect, energy may improve.

Hydrogen and athletic performance

When additional hydrogen is available, mitochondria can produce a greater energy output. John O'Neill, Australian decathlon champion, established that supplying extra hydrogen made it possible for him to dramatically improve endurance and output while experiencing lower blood pulse, minimal fatigue, and little perspiration. The only way O'Neill could explain the energy and power he repeatedly experienced was to suggest a shift from the use of glycogen to the use of hydrogen as a source of fuel.[18] O'Neill referred to this state as HEZ—hydrogen efficiency zone.

O'Neill spent years documenting a regimen that led to HEZ. His routine included large doses of enzyme CoQ10 as a source of hydrogen.[19] Following significant exercise, his body entered a phase where continued workout required minimal exertion. During periods in HEZ, O'Neill was able to achieve greater output with less effort and little, if any, lactic acid buildup. Preliminary studies with hydrogen-rich water reveal a similar reduction in muscle fatigue and lactate buildup for endurance athletes.[20]

Life on Earth evolved in an environment much richer in hydrogen than today. The primordial Earth atmosphere may have contained hydrogen in concentrations as high as 30%.[21] Today, hydrogen is a trace gas, measured in parts per billion. The amount found in the atmosphere is less than 1%. Hydrogen obtained from air, water, and food in today's world is compromised, and it is safe to say that most individuals are hydrogen deficient. No wonder so many symptoms respond favorably to the consumption of hydrogen-rich water.

Oxygen

Ever since Otto Warburg's Nobel Prize–winning work on cellular respiration and his theory about the connection between oxygen and cancer, abundant research has focused on oxygen's role in metabolism. Yet today there are still huge gaps in our understanding. One reason for this is that oxygen is an oxidant; it produces free radicals, which have been viewed as the cause of disease and aging. Yet oxygen-rich water is reported to improve a wide array of symptoms. Ozonated water (an even stronger oxidant) and ozone therapies have an unprecedented record of success. In order to understand this, we need to understand oxidation and the role of oxygen free radicals.

Oxidation

Oxidation is the process that takes place during a fire (a rapid oxidation reaction) as well as the process that causes an apple to turn brown when it is cut (a slow oxidation reaction). Both are examples of breakdown that results when substances lose electrons. Oxidation is Nature's primary method of decomposition. It is also the process used by the immune system to "burn up" invading organisms and neutralize toxins. In water, oxygen is responsible for the decomposition of harmful wastes and the elimination of pathogens. This is accomplished as oxygen radicals steal electrons.

Oxygen radicals

Having two incomplete electron shells, a single oxygen atom is unbalanced and highly reactive. It combines with another oxygen atom to become the stable form of oxygen known as O_2, but it can also be found as a variety of unbalanced free radicals referred to as oxygen radicals—also called reactive oxygen species (ROS) or simply active oxygen. As by-products of metabolism, oxygen radicals can result in oxidative stress (free radical damage) when they are not accompanied by corresponding antioxidants. Research reveals that oxygen radicals can weaken cell membranes, damage DNA, and inactivate many enzymes. This is the reason they have been viewed with disdain. But oxygen radicals are beneficial in many ways too. They play a fundamental role for the immune system. They also participate in cellular communication, regulate growth and gene expression, and are involved in many regulatory processes.[22,23] The healthy human body depends on oxygen radicals. It is the lack of adequate antioxidant buffering capability that is the real problem.[24]

Redox: the balance between oxygen and hydrogen

Oxidation is balanced by a reaction of the opposite nature. When one substance loses electrons (is oxidized), another substance gains electrons (is reduced). This potential for electron exchange is described as redox potential—*red*uction and *ox*idation. Oxygen (the masculine force in water) is the most prevalent oxidizing agent, and hydrogen (the feminine force) is Nature's primary reducing agent (reductant/antioxidant). (Both hydrogen ions (H^+ and H^-) neutralize free radicals.) The combined presence of oxygen radicals and hydrogen-containing reductants help water and the human body to maintain redox balance.

Ozonated water

Ozone is an oxygen radical composed of three oxygen atoms (O_3). It reacts at lightning speed to neutralize most contaminants and harmful microbes in water. Ozone is dipolar, like water, and highly soluble. The addition of ozone to water creates a beautifully structured water/ozone matrix[25] that can hold much more oxygen. However, ozonated water cannot be stored more than several hours. At room temperature, 50% of the ozone is gone in 10 hours. Ozonated water must be made at the time it is used.

When salts are present in water, ozone is even more soluble.[26] In fact, in a saline solution, ozone results in the production of oxygen radicals[27] often referred to as redox signaling molecules.[28] These are abundant in the ocean and in the healthy human body. They are important in maintaining redox balance, and they can support many biochemical and immunological processes.[29] Not only does ozonated saline assist in the eradication of harmful pathogens in water; when consumed, it can increase oxygen utilization and stimulate the production of ATP. It can also stimulate the production of enzymes that act as antioxidant buffers.[30]

Generation of energy and light

Oxygen radicals also generate a source of energy in biological organisms—light.[31] Oxygen radicals contribute to the only biochemical reaction that produces enough energy to generate light.[32] The process is similar to photosynthesis. It has been postulated that oxygen radicals generate the light that creates the unifying photon field characteristic of all life.[33] In other words, the production of oxygen radicals associated with healthy metabolism may contribute to the electromagnetic field surrounding our bodies.

In water, oxygen radicals are continuously generated during dissociation,[34] which is the natural and ongoing splitting of water molecules. Electrons excited by oxygen radicals continue to pass energy from one electron to another until the energy is used for biochemical processes.[35] This is one reason movement is so energizing for water; it brings in oxygen and contributes to dissociation. When the water is structured, there is a well-developed matrix for the ongoing regeneration of excited electrons. However, if water lacks coherent structure, the repeated excitation of electrons is likely to break down more rapidly. This can be illustrated by an electrophoton capture technique that quantifies the amount of light energy (as photon emissions) around the human body[36] and around structured water. Older individuals typically have a reduced photonic field—and less structured water in their bodies. Likewise, water itself that lacks structure has a reduced photonic field (see Appendix A).

Research continues to clarify the relationship between hydrogen and oxygen for the production of energy and free radical balance, and in the maintenance of redox signaling cycles. The presence of these gases and the resulting signaling molecules in water is vital. The presence of carbon dioxide is also vital to maintain the balance between oxygen and hydrogen and to help carry minerals in water.

Carbon dioxide

Carbon dioxide (CO_2) is another trace gas that readily dissolves in water. Even though the atmospheric concentration of this gas is small, it is about 40 times more soluble than oxygen,[37] especially when the water is empty/demineralized. The presence of CO_2 is the reason demineralized water always has an acidic pH (empty water grabs CO_2 from the air and then forms carbonic acid and releases hydrogen ions). Thus, CO_2 in water increases available hydrogen.

$$H_2O + CO_2 \gg H_2CO_3 + H^+$$

Water + Carbon dioxide » Carbonic acid + Hydrogen ion

In 1992, Vladimir Volkov, head of the Laboratory of Biorhythmic Research for the Russian National Institute of Health, proposed that the cause of disease and aging was a reduction of available hydrogen protons (H^+).[38] He also proposed that the condition referred to as acidosis is the body's way to generate H^+ to counteract disease and aging by neutralizing free radicals. (An acid is a substance that releases H^+ ions and thereby reduces free radicals.) Dr. Volkov's research reveals that organisms create acidosis when H^+ is depleted. The acidic condition may therefore be a part of the body's *solution* rather than the source of the problem. When adequate hydrogen is supplied, the body does not need to create an acidic condition. Evidence supports this hypothesis. It is one reason so many hydrogen-based therapies are emerging in medicine.

The two main carriers of hydrogen protons (H^+) are water and carbonic acid. Dr. Volkov asserts that carbonic acid is an organism's best donor of protons. Carbonic acid not only releases H^+ ions, but it also paves the way for pH balance.

pH and buffering capacity

Carbonic acid (H_2CO_3) is a weak organic acid. The bicarbonate ion (HCO_3) is a weak base. Together, these compounds make up the buffering system referred to as the *bicarbonate buffering system* that helps water (and biological tissues) resist dramatic changes in pH. When acids are added to a solution buffered with carbonic acid and bicarbonates, the excess H^+ ions re-associate with bicarbonate ions and re-form carbonic acid. This reduces the H^+ ion concentration (acidity) and brings the pH in check. In nature, carbonic acid, bicarbonate, and carbonate are so interchangeable that they interconvert in a fraction of a second. In water, the presence of one indicates the presence of them all.

$$H_2CO_3 + H_2O \gg HCO_3^- + H_3O^+$$

Carbonic acid + Water » Bicarbonate + Hydronium

Contrary to much publicized information, water with a pH slightly below 7 is not necessarily bad. Many natural springs and rivers have a healthy pH just below 7 due to the water's carbonic acid content. The CO_2–bicarbonate equilibrium gives water the ability to maintain *balance*, which is much more important than pH alone. Water with a slightly acidic pH due to carbonic acid is also capable of holding more minerals. Minerals also help to balance acidity. In solution with carbonic acid, they are more biologically available because of their association with carbon. Technically, these minerals can be called *organic* minerals.

Organically complexed minerals

Carbon is the basis of all organic life. In fact, the term *organic* (as opposed to inorganic) means "containing carbon." Minerals carried with carbon complexes are referred to as organically complexed minerals, or organic minerals. Because of their association with carbon, they are easily recognized by biological organisms and are more rapidly assimilated. The presence of carbon dioxide → carbonic acid → bicarbonates in water provides a way for minerals to be organically complexed and thus more biologically available.

Another way minerals are carried organically is with a group of organic acids called fulvic acids. Fulvic acids (hundreds have been identified) are exuded from microbes during the breakdown of organic matter in the soil. Their elemental composition (carbon, hydrogen, oxygen, nitrogen, and sulfur) includes the same basic elements that make up all life. Fulvic acids are simple and at the same time, extremely complex. Their structure includes multiple voids and binding sites for carrying minerals and other plant complexes. They are completely soluble in water—even when holding many times their weight in minerals. When fulvic acids are added to water, inorganic minerals are assimilated and carried within the organic complex. They become biochemically active and biologically available.

Minerals in water were meant to be carried along with organic acids such as carbonic acid and fulvic acids. Unfortunately, these are routinely removed prior to chlorination because they form toxic by-products called trihalomethanes (THMs) when mixed with chlorine. The use of chlorine therefore destroys Nature's way of providing minerals in their natural, organic form. After water treatment, minerals are left as *in*organic minerals.

All natural, unprocessed salts contain carbon complexes and bicarbonates—although to a small degree. By comparison, seawater has a relatively high concentration of bicarbonates and fulvic acids. The addition of salts, seawater, fulvic acids, carbon dioxide, or even a slice of lemon adds organic complexes to water that help balance the inorganic minerals. The presence of carbon is the key. The addition of CO_2 also assists in the formation of another form of therapeutic water—bicarbonate water.

Adding carbonates and bicarbonates to water

Carbonated water by itself can have a pH between 3 and 5 (about the same as apple or orange juice). But when minerals are present, a wide variety of carbonates and bicarbonates form (including calcium bicarbonate, magnesium bicarbonate, sodium bicarbonate, potassium bicarbonate, and others), and the pH becomes more balanced. Bicarbonates have a variety of therapeutic benefits.

For well over a hundred years, sodium bicarbonate (baking soda) has been used to increase the buffering capacity of the blood and the fluid surrounding cells.[39] It has many documented uses in emergency medicine (although its use is being rapidly phased out in favor of drugs). Dr. Mark Sircus has been a proponent of sodium bicarbonate for many years. His 2014 book elaborates on its chemistry and function for human health:[40]

> Sodium bicarbonate is the time honored method to "speed up" the return of the body's bicarbonate levels to normal. It has blood vessel dilating action, [it] increases blood fluidity, assist[s] oxygen dissociation—thus more oxygen flows to the capillaries and cells . . . has strong anti-inflammatory action, helps with detoxification and neutralization of toxic substances of all kinds offering strong and almost instantaneous shifts in pH. . . . The body is always hungry for bicarbonate unless you live in some pristine valley.

The consumption of water with bicarbonates on an empty stomach actually stimulates the production of stomach acid (HCl) since bicarbonates provide the necessary CO_2 required to begin the process.[41] It is an ideal drink first thing in the morning and before meals—as long as adequate time passes before eating. Bicarbonate water prepares the stomach for digestion and helps to overcome the waning production of HCl that plagues older individuals. Even though this knowledge has nearly disappeared from modern medical texts, the *Materia Medica* (1918 edition) has this to say about the effects of sodium bicarbonate in the stomach:

> The effect of an alkali in the stomach will vary according to the nature of the stomach contents at the time of administration. In the resting period (after food is digested) sodium bicarbonate dissolves mucus and is absorbed as bicarbonate into the blood, to increase its alkalinity directly. In the digestive period it reduces the secretion of gastric juice, neutralizes a portion of the hydrochloric acid, liberates the carminative carbon dioxide gas, and is absorbed as sodium chloride. The time of administration must, therefore, be chosen with a definite purpose. In continuous hyperacidity and in fermentative conditions a dose an hour before meals will tend to prepare the stomach for the next meal.[42]

Although sodium bicarbonate is a simple, time-honored favorite, other bicarbonates may be even more beneficial.

Magnesium bicarbonate water

In 2002, Dr. Russell Beckett, an Australian biochemist and pathologist, made an interesting discovery while doing research on anti-aging. He identified a group of animals in a remote area of southeastern Australia living 30 to 50% longer than other animals in the same region. After eliminating all other possibilities, the water was found to have an unusually high level of magnesium bicarbonate.

Beckett's earlier research revealed that animals that broke the *longevity rules* had evolved more efficient ways of neutralizing acid waste.[43] He recognized the importance of bicarbonates in the water, and at the same time he recognized the importance of magnesium. Each component assists in the transport of the other within cellular tissues. Subsequent research proved he was on to something.

Over a period of two years, individuals who drank water with augmented levels of magnesium bicarbonate showed reduced incidence of colds and flu with less severe symptoms.[44] (Most viruses, including the ones that are responsible for the common cold and flu, are classified as *pH dependent*. They are infectious only in an acidic pH range.[45]) Later clinical trials revealed positive changes in the biochemistry of the blood (improved albumin levels) and positive effects on the parathyroid hormone, which is correlated with osteoarthritis, osteoporosis, and atherosclerosis.[46] Supplying magnesium and bicarbonate together in drinking water supports the bicarbonate buffering system in the body. It provides available hydrogen and supplies magnesium in a biologically available form.

Magnesium absorption from water

Unlike many other nutrients that are better supplied in food, water can be an excellent source of magnesium,[47] especially when the minerals are in organic form. According to studies, approximately 50% of the magnesium contained in natural mineral water is absorbed.[48,49] This is greater than the amount absorbed from food.[50] The authors have experienced benefits from water with magnesium bicarbonate that are over and above the benefits from a wide array of magnesium supplements. As far as they are aware, no studies have been conducted to determine the absorption of magnesium from magnesium bicarbonate water; they suspect it is much higher than 50%.

Blatantly ignored in modern medicine, magnesium deficiency is epidemic. It is responsible (at least in part) for the increased incidence of heart disease, stroke, diabetes, asthma, arthritis, high blood pressure, chronic fatigue syndrome, attention deficit, and many other conditions.[51] Magnesium has been referred to as the "master mineral" because of its involvement in hundreds of enzyme systems. Interestingly, it is required in the mitochondria along with hydrogen and oxygen for the production of ATP. According to some authorities, total magnesium intake should be at least 450 to 500 mg per day, and water should contain a minimum of 25 to 50 mg/l.[52] The magnesium concentration in the ocean is over twice the concentration of calcium. This is ideal for all water. Unfortunately, most freshwater today has too much calcium and hardly any magnesium. The majority of today's water contains under 6 mg/l of magnesium.[53] Magnesium bicarbonate water is an excellent way to receive the benefits of organically complexed magnesium *and* bicarbonates.

Alkaline water

The alkaline minerals (Ca^{2+}, Mg^{2+}, Na^{+}, K^{+}) in alkaline ionized water and in many alkaline water supplements are carried as hydroxides, attached to the hydroxide ion (OH^{-}). In this inorganic form, these minerals can contribute to degenerative diseases.[54] However, when the same alkaline ions are carried with bicarbonates or otherwise organically complexed, they are biologically available and free to participate with the bicarbonate buffering system that actually keeps arterial plaque and age-related calcification in check.[55] The presence of organic acids (including fulvic acids and carbonic acid) allows water to hold more minerals—in organic form. In this form, they are easily assimilated; they balance the water and help to buffer the pH.

The influence of temperature on dissolved gases

Temperature is a big factor in water's ability to dissolve gases. In warm water, there is greater molecular motion, and gas molecules tend to gather together, forming bubbles that eventually rise to the surface and escape. Cold water holds more dissolved gases than warm water. It has a more stable molecular structure and forms organized layers around gases. This keeps them in a less reactive state while they become a part of the liquid crystalline matrix.

In Part II you will learn how to make hydrogen-rich water, oxygen/ozone-rich water, and magnesium bicarbonate water.

References (Chapter 5)

1 Szent-Györgyi, A. (1937). Oxidation, Energy Transfer and Vitamins, Nobel lecture.

2 Dohi, K et al. (2014). Molecular Hydrogen in Drinking Water Protects against Neurodegenerative Changes Induced by Traumatic Brain Injury. PLoS One, 9(9). Available online: http://www.ncbi.nlm.nih.gov/pmc/articles/PMC4176020/

3 Batmanjheligj, F. (2003). *Water: You're Not Sick, You're Thirsty*, Warner Books, pp. 183-187.

4 Huang, C. Kawamura, T. Toyoda, Y. and Nakao, A. (2010). Recent Advances in Hydrogen Research as a Therapeutic Medical Gas. *Free Radical Research* 44(1), pp. 971-982.

5 Zhang, J. et al. (2012). A Review of Hydrogen as a New Medical Therapy. *Hepato-Gastroenterology* 59(1), pp. 1026-1032.

6 Ohta, S. (2012). Molecular Hydrogen Is a Novel Antioxidant to Efficiently Reduce Oxidative Stress with Potential for the Improvement of Mitochondrial Diseases. *Biochimica et Biophysica Acta* 1820, pp. 586-94.

7 Nagata, K. et al. (2009). Consumption of Molecular Hydrogen Prevents the Stress-induced Impairments in Hippocampus-dependant Learning Tasks During Chronic Physical Restraint in Mice, *Neuropsychopharmacology*, 32(2), pp. 501-508. Available online: http://www.nature.com/npp/journal/v34/n2/abs/npp200895a.html.

8 Hong, Y, Chen, S. and Zhang. J-M. (2010). Hydrogen as a Selective Antioxidant: A Review of Clinical and Experimental Studies. *Journal of International Medical Research* 38: 1893-1903. Available online: http://imr.sagepub.com/content/38/6/1893

9 Dixon, B. Tang, J. and Zhang, J. (2013). The Evolution of Molecular Hydrogen: a noteworthy potential therapy with clinical significance. *Medical Gas Research*, 3(10). Available online: http://www.medicalgasresearch.com/content/3/1/

10 Zhao L. Zhou C. Zhang J. Gao F. Li B. Chuai, Y. Liu, C. and Cai, J: (2011). Hydrogen Protects Nice From Radiation Induced Thymic Lymphoma in BALB/c Mice. *International Journal of Biological Science*, 7(10), pp. 297-300.

11 Giambarresi, I. and Walker, R. *Medical Consequences of Nuclear Warfare*, Chapter 11: Prospects for Radioprotection. 285. Available online: http://www.dancingwithwater.com/wp-content/uploads/HydrogenforRadiation-USMilitarydoc.pdf.

12 Kamimura, N. Nishimaki, K. Ohsawa, J. and Ohta, S. (2011). Molecular Hydrogen Improves Obesity and Diabetes by Inducing Hepatic FGF21 and Stimulating Energy Metabolism in db/db mice. *Obesity (Silver Spring)*, 19(7), pp. 1396-1403.

13 Aoki, K. Nakao, A. Adachi, T. Matsui, Y. and Miyakawa, S. (2012). Effects of Drinking Hydrogen-rich Water on Muscle Fatigue Caused by Acute Exercise in Elite Athletes. *Medical Gas Research,* 2(12). Available online: http://www.ncbi.nlm.nih.gov/pmc/articles/PMC3395574/

14 Kajiyama, S. et al. (2008). Supplementation of Hydrogen-rich Water Improves Lipid and Glucose Metabolism in Patients with Type 2 Diabetes or Impaired Glucose Tolerance. *Nutrition Research*, 28(1), pp. 137-143.

15 Dohi, K et al. (2014). Molecular Hydrogen in Drinking Water Protects against Neurodegenerative Changes Induced by Traumatic Brain Injury. *PLoS One* 9(9). Available online: http://www.ncbi.nlm.nih.gov/pmc/articles/PMC4176020/

16 Hong, Y. Chen, S. and Zhang. J-M. (2010). Hydrogen as a Selective Antioxidant: A Review of Clinical and Experimental Studies. *Journal of International Medical Research*,38(1), pp. 1893-1903. Available online: http://imr.sagepub.com/content/38/6/1893

17 Szent-Györgyi, A. (1937). Oxidation, Energy Transfer and Vitamins. Nobel lecture.

18 O'Neill, J. (1994). *Breaking the Sound Barrier of Human Energy Production*, Published by John O'Neill.

19 O'Neill, J. (1994). *Breaking the Sound Barrier of Human Energy Production*

20 Aoki, K. Nakao, A. Adachi, T. Matsui, Y. and Miyakawa, S. (2012). Effects of Drinking Hydrogen-rich Water on Muscle Fatigue Caused by Acute Exercise in Elite Athletes. Medical Gas Research, 2(12). Available online: http://www.ncbi.nlm.nih.gov/pmc/articles/PMC3395574/

21 Tian, F. et al. (2005). A Hydrogen-Rich Early Earth Atmosphere, *Science*, 308(5724), pp. 1014-1017. Available online: http://www.sciencemag.org/cgi/content/abstract/308/5724/1014

22 Gitler, C. and Danon, A. editors. (2003). *Cellular Implications of Redox Signaling*, World Scientific.

23 Sarsour, E. et al. (2009). Redox Control of the Cell Cycle in Health and Disease, Antioxidants and Redox Signaling, 11(12), pp. 2985-3010.

24 Shallenberger, F. (2011). *The Principles and Applications of Ozone Therapy: A Practical Guide for Physicians,* p. 18.

25 Shallenberger, F. (2011). *The Principles and Applications of Ozone Therapy: A Practical Guide for Physicians,* p. 108.

26 Shallenberger, F. (2011). *The Principles and Applications of Ozone Therapy: A Practical Guide for Physicians,* p. 108.

27 Bocci, V. Valacchi, G. Corradeschi, F. Aldinucci, C. Silvestri, S. Paccagnini, E. and Gerli, R. (1998). Studies on the Biological Effects of Ozone: 7. Generation of

reactive oxygen species (ROS) after exposure of human blood to ozone. *Journal of Biological Regulators and Homeostatic Agents*, 12(3):67-75.

28 Samuelson, G. (2009). *The Science of Healing Revealed: New Insights into Redox Signaling*. p. 46.

29 Samuelson, G. (2009). *The Science of Healing Revealed: New Insights into Redox Signaling*.

30 Elvis , A. and. Ekta, J. (2011). Ozone Therapy: A Clinical Review. *Journal of Natural Science, Biology and Medicine*, 2(1): 66–70.

31 Voeikov, V. (2006). Reactive Oxygen Species (ROS): pathogens or sources of vital energy? Part 2. Bioenergetic and bioinformational functions of ROS. *Journal of Alternative and Complementary Medicine*. 12(3), pp. 265-270.

32 Taylor, R. (2006). Free Radicals and the Wholeness of the Organism, *NEXUS Magazine*,13(3).

33 Voeikov, V. (2001). Reactive Oxygen Species, Water, Photons and Life, Rivista di Biologia, Biology Forum 94, pp. 193-214. Available online: http://gdv.reimei.tv/pdf2.pdf

34 Voeikov, V. (2001). Reactive Oxygen Species, Water, Photons and Life, Rivista di Biologia, Biology Forum 94, pp. 193-214. Available online: http://gdv.reimei.tv/pdf2.pdf

35 Voeikov, V. (2001). Reactive Oxygen Species, Water, Photons and Life, Rivista di Biologia, Biology Forum 94, pp. 193-214. Available online: http://gdv.reimei.tv/pdf2.pdf

36 Taylor, R. (2006). Free Radicals and the Wholeness of the Organism, NEXUS Magazine,13(3).

37 Website Engineering Toolbox http://www.engineeringtoolbox.com/gases-solubility-water-d_1148.html

38 Volkov, V. Hydrogen Against Aging, Illnesses and Death. Available online: http://www.health-freedom.info/hplus/c18m3t1.html

39 Bastedo, W. (1918). *Materia Medica: Pharmacology: Therapeutics and Prescription Writing*. 2nd edition. W.B. Saunders Company, pp. 94-95. Available online: https://archive.org/stream/materiamedicapha00bastuoft#page/n1/mode/2up

40 Sircus, M. (2014). *Sodium Bicarbonate: Nature's Unique First Aid Remedy*, Square One Publishers.

41 Science Student Center Website *Biology*, 8th edition Chapter 48: The Digestive System. Raven Available online: http://highered.mheducation.com/sites/9834092339/student_view0/chapter48/hydrochloric_acid_production_of_the_stomach.html

42 Bastedo, W. (1918). *Materia Medica: Pharmacology: Therapeutics and Prescription Writing. 2nd edition.* W.B. Saunders Company, pp. 94-95. Available online: https://archive.org/stream/materiamedicapha00bastuoft#page/n1/mode/2up

43 Bowers, P. (2002). Peter Bowers on the Clues that Led to the Water. Online article http://www.smh.com.au/articles/2002/04/09/1018302715172.html

44 Beckett, R. US Patent #6,048,553 Aqueous Metal Bicarbonate Solution Useful in Treating Inflammatory, Degenerative and Viral Diseases.

45 Sircus, M. (2014). Viral Infections are pH Sensitive, Available online. http://drsircus.com/medicine/viral-spread-ph-sensitive

46 Today/Tonight News exclusive https://www.youtube.com/watch?v=-kVFjQ-ZrEs Unique Water.

47 Marier J. (1981). Nutritional and Myocardial Aspects of Magnesium in Drinking Water. *Magnesium Bulletin* 48–54.

48 Sabatier, M. et al. (2002). Meal Effect on Magnesium Bioavailability from Mineral Water in Healthy Women. *American Journal of Clinical Nutrition*, 75(1), pp. 65-71.

49 Verhas, M. et al. (2002). Magnesium Bioavailability from Mineral Water: a study in adult men. *European Journal of Clinical Nutrition,* 56(1), pp. 442-47.

50 Rude R. (2010). Magnesium. In: *Encyclopedia of Dietary Supplements*, Coates, P. Betz, J. Blackman, M. Cragg, G. Levine, M. Moss. J. White. J. eds. 2nd edition. Informa Healthcare, pp. 527-37.

51 The Magnesium Online Library Website: http://www.mgwater.com/index.shtml

52 Altura, B. (2009). Atherosclerosis and Magnesium. In: WHO: Calcium and Magnesium in Drinking-Water: Public Health Significance. Geneva: WHO, pp. 75–81.

53 Azoulay, A. Garzon, P. Eisenberg, M. (2001). Comparison of the Mineral Content of Tap Water and Bottled Waters. *Journal of General Internal Medicine*, 16(1), pp. 168–175.

54 Xiao, Q. Murphy, R. Houston, D. Harris, T. Chow, W-H. Park, Y. (2013). Dietary and Supplemental Calcium Intake and Cardiovascular Disease Mortality. The National Institutes of Health–AARP Diet and Health Study, *JAMA Internal Medicine,* 173(8). Available online: http://archinte.jamanetwork.com/article.aspx?articleid=1568523

55 Leibrock, C. Voelkl, J. Kohlhofer, U. Quintanilla-Martinez, L. Kuro-o, K. and Lang, F. (2016). Bicarbonate-Sensitive Calcification and Life Span of Klotho-deficient Mice. *American Journal of Physiology: Renal Physiology*, 310(1). Available online: http://ajprenal.physiology.org/content/310/1/F102

Chapter 6
Geometry

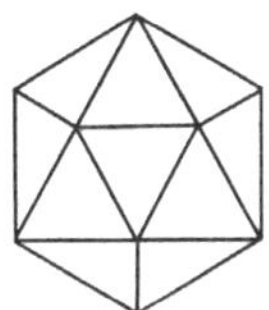

Bringing structure to water requires energy.[1] Many forms of energetic input will satisfy this requirement; however, sources that are tuned to the geometry and natural harmonics of the organic universe produce more lasting structure. Unnatural forms (in other words, those not found in Nature) do not produce the coherence required for lasting molecular organization. This chapter discusses the influence of geometry for structuring water.

The source field

The field of energy in which we reside has been referred to by a variety of different names. It has been called the aether, the zero point field, the vacuum, and even dark energy. Today, some refer to it as the *source field* or the universal field of energy. Regardless of name, the universe is teeming with energy. Galaxies, stars, planets, and all biological organisms continually interact with it, extracting and translating the energy it holds (see Torsion fields in Chapter 3).

One of the reasons *space* has escaped our investigation as the ultimate source of energy is that we are immersed in it. Even though the energy of space provides sustenance in ways we do not yet fully comprehend, the extractive process is passive—similar to that of breathing. Many are coming to understand that if we were cut off from its influence we would die, regardless of how much air, food, or water are available. Our own bodies continuously gather energy from the universal source field. The terms chi, ki, and prana describe the translation of this energy into what might also be called *organic light* or *life force*.

The geometry of space

The new sciences are exploring the idea that the universal source field has fluid-like characteristics similar to those of water.[2,3,4,5] Investigation reveals that space is likely in continuous spiraling motion. This concept was introduced by Paul Dirac (1902–1984) in conjunction with his Nobel Prize–winning theory of the spin of electrons. He suggested that nothing could spin without *wrapping itself up* in the background medium—unless the medium was in rotational movement. Dirac proposed that the spin of an electron was intimately related to the *structure* of space.[6]

In 2011, work at UCLA determined that the honeycomb-like lattice of graphene (a single layer of graphite) caused electrons to spin in a distinct way. Researchers concluded that *spin . . . can derive from hidden substructure, not of the particles themselves, but rather of the space in which these particles live.*[7] If there is hidden structure in space, then water's tendency toward spiral movement may be the result of the geometry of the source field—as well as from water's own molecular geometry. One way or another, without geometry, there would be no organized pathway for the flow of energy.

The significance of the golden ratio

At the very foundation of geometry is the golden ratio, also referred to as the *golden mean* and *divine proportion.* It is a mathematical ratio that expresses the relationship of two parts with each other and

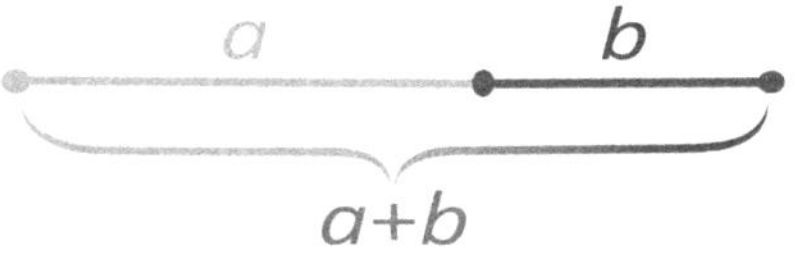

with the whole. For example, there are many places where you could divide a line into two parts. Each part would result in a different ratio for the length of the small to the larger line. However, there is one unique point at which the ratio of the small portion to the large portion is exactly the same as the ratio of the large portion to the whole. This is the golden ratio. It is represented by the Greek letter phi (Φ). Phi is an irrational number (typically rounded to 1.618) with no repeating pattern. Although it is irrational, the golden ratio is said to be the most *perfectly* irrational number there is. With an *infinite number of harmonic resonance points*,[8] the golden ratio resonates with the entire repertoire of creative forces in the universe. It represents a hidden harmony in the natural world.

The golden ratio has fascinated mathematicians and scientists for centuries. It establishes visual balance that has become a standard for width in relation to height in design and architecture. The golden ratio also describes the growth of some of Nature's most complex and orderly systems. For example, golden ratio proportion describes the dimensions of insects and animals, including things like the proportions of the human face, body, fingers, and teeth. The golden ratio also appears in the proportions of basic shapes, as well as in more complex 3-D solids. You could say the golden ratio is the *alphabet* in the language of creation.

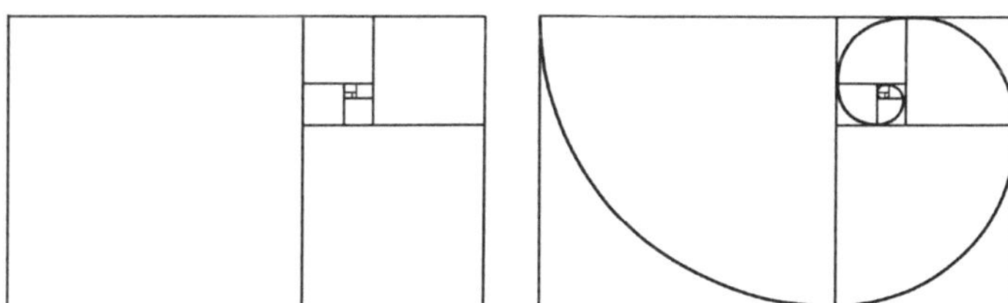

The golden ratio creates a golden rectangle (above left). When a square is sectioned off from the rectangle, a smaller rectangle of the same proportion remains. The pattern repeats infinitely. Curved lines drawn within each square create a perfect quarter circle; they

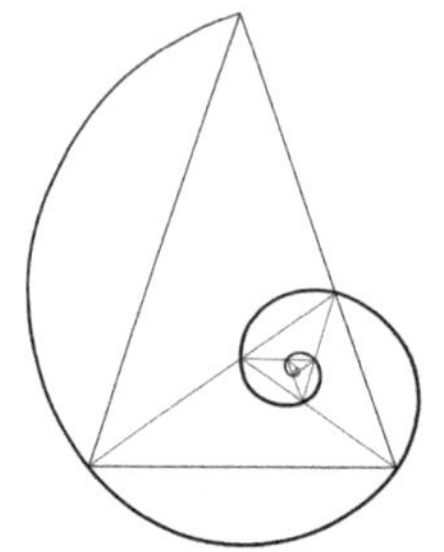

contribute to the generation of a golden ratio spiral, where the *whole* increases by a factor of phi at each quarter turn. A similar pattern exists in the golden triangle (left), where each portion of the resulting spiral represents one-third of a circle and the size of the whole increases by a factor of phi at each one-third turn.

The golden ratio spiral is a logarithmic fractal,* where each part is a smaller version of the whole. It is related to the phenomenon of phylotaxis—the tendency for growth to occur in spiral patterns. Phylotaxis can be found everywhere in the realm of things that unfold in steps. It represents the most efficient pattern in which growth can occur, from the pattern of the seeds in a sunflower to the spiraling pattern of a pinecone and a seashell. Golden ratio spirals may focus and amplify source energy, helping to bring organic light/life force into physical form.

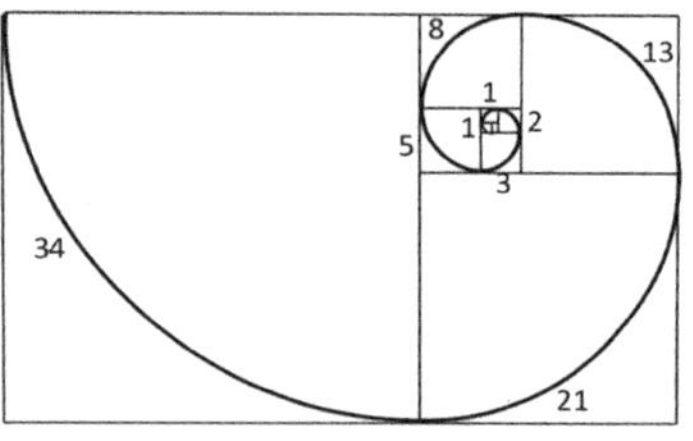

Closely related to the golden ratio is the Fibonacci sequence—a series of numbers that increase by adding the last number to the previous number: 0, 1, 1, 2, 3, 5, 8, 13, 21, 34 and so on. If you divide the last number in the sequence by the previous one, you get a number close to phi. The further you go in the series, the closer the result is to phi. The Fibonacci sequence is a way to approximate the golden ratio using whole numbers. It is used in many man-made devices, but it is not a substitute for phi.

Platonic solids

As far back as the Greek Mystery Schools, five perfect 3-D forms have been recognized as the foundation of everything in the physical world. These forms, known as the Platonic solids, also have their

* The term "fractal" refers to self-similarity on all levels; it is a repeated geometric pattern that can be scaled to any size.

roots in golden ratio proportion. In 2-D space, regular geometrical figures are triangles, squares, pentagons, etc. When fitted together, some of these make 3-D figures: tetrahedron (4 triangles), cube (4 squares), octahedron (8 triangles), dodecahedron (6 pentagons), and icosahedron (20 triangles). These are the only five solids that have the same shape on every side and only one angle per edge; if you spin any of the Platonic solids around its center point, the corners would create a perfect sphere.

Modern scholars dismissed the idea that Platonic geometry could be at the foundation of our universe until the 1980s, when Professor Robert Moon (1911–1989), one of the pioneers of nuclear energy, demonstrated that the Periodic Table of Elements is based on the Platonic solids.[9] Nesting of the Platonic solids is the basis for atomic and molecular structure. The Platonic solids also play a role in chemistry as molecular organizational structures, and they are the basis for crystal patterns that occur throughout the mineral kingdom.

Platonic solids are the building blocks of Nature. They form the basic units of matter and energy. You could say they are the *words* to the language of creation. No wonder geometry is apparent in the study of mathematics, music, cosmology, and all the life sciences.

Flower of Life: life force matrix

One of the best illustrations of the Platonic solids as building blocks is found in an ancient motif called the Flower of Life. This fractal pattern, built using intersecting circles where each circle's circumference crosses other circles' center points, contains each of the Platonic solids, as illustrated on the next page.

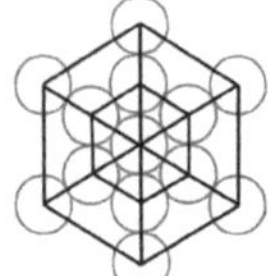 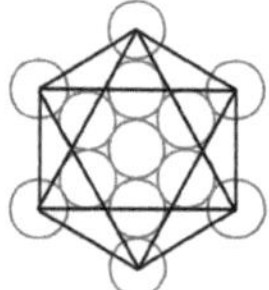 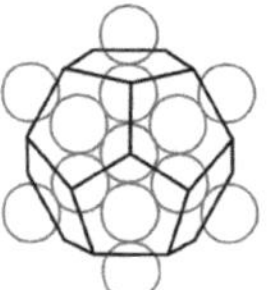 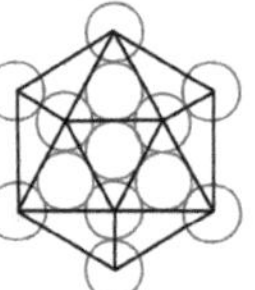 

Found within the geometry of the Flower of Life, from left to right, are the hexahedron (cube), octahedron, dodecahedron, icosahedron, and star tetrahedron.

Inherent in the geometry of the Flower of Life is a hexagonal pattern derived from six-fold symmetry. The same pattern is found on a larger scale throughout the universe. The following graphics illustrate the fractal nature of the Flower of Life, where the same pattern is used to create ever-increasing complexity. It is no coincidence that water's structure is also based on hexagonal symmetry and no wonder that water has the capacity to hold all the patterns of the universe.

Three developmental levels of the Flower of Life illustrate the creative potential from one simple pattern. Shown, from left to right, are the Seed of Life, the Flower of Life, and the Complete Flower of Life.
The Complete Flower of Life was conceived of and created by Andrew Monkman.

Shapes as wave guides

There are many shapes that echo golden ratio proportion. According to research conducted by Dan Davidson, the unique characteristics of pyramids are a function of their angles—each creating a vortex (wave guide) for source energy.[10] Davidson and others have shown that pyramids generate very mild magnetic fields that increase as the orientation approaches a north-south/east-west alignment.[11] Davidson concluded that pyramids are transducers of source energy, bringing it into electromagnetic form. As such, they can have an influence on water's structure.

Russian researchers came to a similar conclusion. Large golden ratio pyramids constructed in Russia in the 1990s were shown to keep distilled water in a liquid state—even at temperatures many degrees below freezing (–38° C/–6° F). This phenomenon is characteristic of highly structured water.[12] Diamonds synthesized inside the Russian pyramids were harder and more pure than diamonds synthesized elsewhere.[13] This is further evidence of the supportive nature of pyramidal geometry for the development of crystalline order.

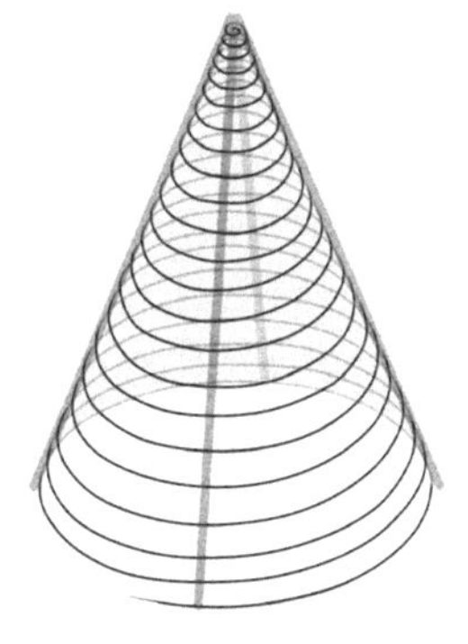

The shape of the cone is another powerful wave guide for source energy. In many ways, it is an infinite-sided pyramid, and it functions similarly.[14] Research conducted at the Physics Institute of the National Academy of Sciences of Ukraine and at Chernovetsky University concluded that cones constructed according to golden ratio geometry were strongly supportive of many natural processes.[15] The Native American teepee is a good example of a conic shape used to focus energy in the living environment. As with pyramids, cones capture and translate source energy. They can be highly supportive when placed around water.

The shape of the egg

The egg is a primordial structure also based on golden ratio proportion. It is the 3-D geometry of creation and the ideal shape for water to be placed *within.* Epitomized by an infinite number of 2-D golden ratio spirals (shown), an egg focuses source energy. The shape of the egg encourages gentle, cyclic movement ideal for

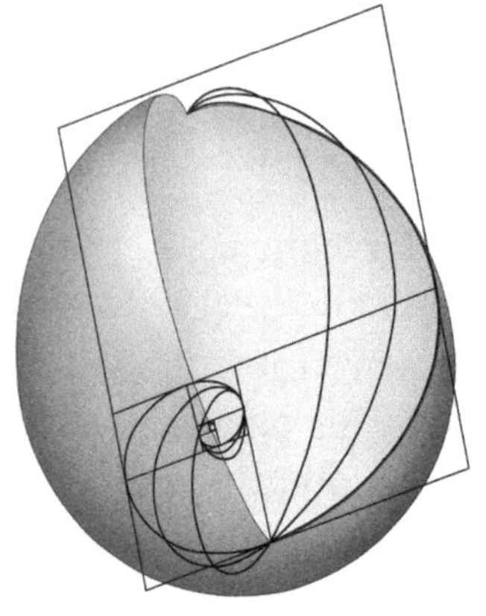

the development of water's coherent structure. Water can remain within an egg-shaped vessel for a long time without becoming "stale." The longer it remains, the more refined water becomes.

The entire universe is based on geometry that encourages spiraling motion. Movement creates torsion fields (vortices) that conduct the flow of energy and information. According to Davidson, and others, source energy can be directed, focused, and intensified by shapes, geometry (including the geometry of crystals), magnetic fields, and even thought. Water responds to energy in all these forms and is especially responsive to the toroidal energy field surrounding biological organisms. As the basis for life, geometry can be used in the creation of organic, living systems. It has also been used by the dark arts for entrapment and artificial life. The intention of the creator makes the difference.

In the next several chapters, we will discuss the energetic effects of light, vibration (sound), electricity, magnetism, and focused attention on water's molecular structure.

Image credits (Chapter 6)

Page 74 Complete Flower of Life created by Andrew Monkman
www.newparadigmjournal.com/Oct2008/flower.htm

References (Chapter 6)

1 Pollack, Gerald. (2013). *The Fourth Phase of Water.* Ebner & Sons, pp. 87-95.

2 Haramein, N. Rauscher, E. and Hyson, M. (2008). Scale Unification: a universal scaling law for organized matter. *Proceedings of the Unified*

Theories Conference. ISBN 9780967868776

3 Wilcock, D. *Divine Cosmos* online book available at http://divinecosmos.com

4 Davidson, D. (1997). *Shape Power: A Treatise on How Form Converts Universal Aether into Electromagnetic and Gravity Forces and Related Discoveries in Gravitational Physics.* Rivas Publishers.

5 Huong, K. (2013). Dark Energy and Dark Matter in a Superfluid Universe. Invited talk at the Institute of Advanced Studies, Nanyang Technology University, Singapore Aug 26-29, Available online: http://arxiv.org/ftp/arxiv/papers/1309/1309.5707.pdf

6 Batty-Pratt. E. and Racey, T. (1980). Geometric Model for Fundamental Particles. *International Journal of Theoretical. Physics,* 19(1), pp. 437-475.

7 Mecklenburg, M. and Regan, B. (2011). Spin and the Honeycomb Lattice: Lessons from Graphene. *Physical Review Letters,* 106 (116803) Abstract online: http://journals.aps.org/prl/abstract/10.1103/PhysRevLett.106.116803

8 Davidson, D. (1997). *Shape Power: A Treatise on How Form Converts Universal Aether into Electromagnetic and Gravity Forces and Related Discoveries in Gravitational Physics,* section 4.3. Rivas Publishers.

9 Hetch, L. and Stevens, C. (2004). New Explorations with the Moon Model, *21stCentury Science & Technology Magazine.* Pp. 58-73. Available online. http://www.21stcenturysciencetech.com/Articles%202005/MoonModel_F04.pdf

10 Davidson, D. (1997). *Shape Power: A Treatise on How Form Converts Universal Aether into Electromagnetic and Gravity Forces and Related Discoveries in Gravitational Physics,* section 5.4. Rivas Publishers.

11 Davidson, Dan *Shape Power:* A Treatise on How Form Converts Universal Aether into Electromagnetic and Gravity Forces and Related Discoveries in Gravitational Physics section 4.8.5 Rivas Publishers.

12 Jhon, M-S. (2004). *The Water Puzzle and the Hexagonal Key,* 2004 Uplifting Press, pgs 34-35.

13 Pyramids.RU Website: http://www.pyramids.ru/

14 Davidson, D. (1997). *Shape Power: A Treatise on How Form Converts Universal Aether into Electromagnetic and Gravity Forces and Related Discoveries in Gravitational Physics,* Introduction.

15 Wilcock, D. *Divine Cosmos* (chapter 9.5) online book available at http://divinecosmos.com

Chapter 7

The Influence of Light

Sunlight delivers life-supporting energy and information to all life forms on the planet. It can also satisfy the *energy requirement* for the development of structured water. Yet sunlight can be both destructive and constructive of water's liquid crystalline lattice, depending on its wavelength and intensity and the duration of exposure. Some wavelengths cause water molecules to dissociate, and others contribute to the development of lasting, liquid crystalline structure. As with everything in nature, all wavelengths play a valuable role.

Destructive and constructive wavelengths

Light from the Sun is a mixture of electromagnetic wavelengths that range from infrared (IR) to ultraviolet (UV), with the visible spectrum in between. The Sun also emits X-rays and gamma rays that never reach the Earth, instead being absorbed in the atmosphere. The most destructive UV light is also absorbed before it reaches the Earth.

Although only a small portion of UV light is allowed to penetrate to the surface of the Earth, it can be disruptive to almost all biological life forms. UV light has an electrical component that tends to cause dissociation among water molecules. These fast, short wavelengths break hydrogen bonds, leaving water free for reorganization. IR wavelengths are longer and slower than UV wavelengths. According to research conducted at the University of Washington, IR light increases the organized water zone by dozens of times.[1] Thus, sunlight can provide the energy to structure water—as long as constructive wavelengths (IR) outweigh destructive wavelengths (UV).

Radiation from the Sun includes ultraviolet (UV) rays and infrared (IR) rays on either side of the visible spectrum. UV rays are short, high-frequency vibrations that cause water molecules to dissociate; IR rays are characterized by lower frequencies and longer wavelengths and are supportive of water's developing structure.

Light absorption by water

Sometimes UV and IR wavelengths are not considered "light" because they don't belong to the *visible* spectrum. Yet even though they are unseen, these wavelengths are felt by everything on the planet. UV light is energetically potent and easily absorbed by water (see graph next page). IR light is also readily absorbed by water—and by everything it touches.

IR and far infrared (FIR) rays have been referred to as the *frequencies of life*. IR radiation is everywhere, and it supports all biological life forms. These rays comprise almost 50% of the Sun's energy, providing much of the heat for the planet. They are absorbed and then radiated for a long period of time, even in darkness (think of night vision cameras that pick up IR rays showing images of buildings and life forms in complete darkness). The Earth also continuously radiates IR energy from its central core. IR radiation is a large contributor to water's structure. Its greatest effects are observed during the still, quiet hours of the night, when water reaches the height of its maturity and energetic potential in complete darkness.

The visible light spectrum is a denser form of light perceived by our senses in a matter-based world. It is interesting that visible light is not as well absorbed by water (see graph below). Blue and violet wavelengths are the most difficult for water to absorb. These wavelengths penetrate deeply into water before they can be absorbed. This is the biggest reason deep water appears blue. It is also the reason water's structure is best preserved in blue and deep violet glass. These colors have the least vibratory impact on water. Their influence is similar to that of darkness.

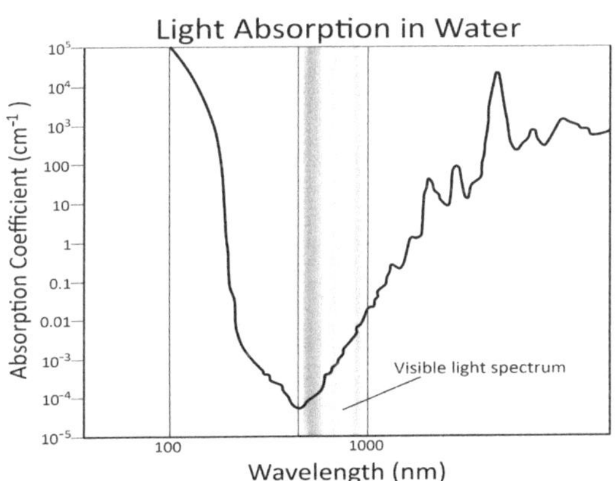

Water's Absorption of Solar Light

UV and IR rays are easily absorbed, whereas violet and blue rays (from the visible spectrum) are not easily absorbed.

Although there are many studies on light absorption in the ocean and in pure water, little attention has been given to how light is absorbed in natural water (for example, springwater, glacial melt, rain, etc.). Most studies to date have been performed using ultrapure water. This kind of water is "sterile" from an energetic perspective. When studies using natural water finally emerge, they will reveal many secrets regarding water's interaction with light and other sources of energy. For example, water with minerals (as opposed to ultrapure water) is known to absorb and hold more light, which is one reason the ocean is teaming with energy and information.

Almost no interest has been expressed in the effects of water's *structure* on the absorption of light; however, it has been reported that water stored in quartz containers exhibits increased absorption in the blue and UV range.[2] We suspect the reason for this is that quartz introduces piezoelectric currents and templating effects (see Crystals page 150) that bring structure to water and increase the absorption of light in the measured range. Greater structure augments many of water's characteristics—including its ability to absorb light and other forms of energy.

Too much Sun

Even though sunlight has an overall structuring effect on water, your drinking water should never be left in the sun. Water left standing in direct sunlight cannot maintain crystalline structure for several reasons:

1. Shallow, standing water has no place to *hide* from the overstimulation by visible and UV light. More than a few minutes in direct sunlight depletes water's energy and destabilizes its structure.
2. Deep water provides darkness and protection from too much light. Heat accumulates in shallow, standing water, causing the organized molecular matrix to break down. Water's optimal temperature is 4° C (39° F). At this temperature it is known to maintain the most stable structure.
3. Standing water is not able to *move* to stay energized. Water in motion (even in the sun) creates vortices that keep it cool and help to gather energy and maintain liquid crystalline order.

Viktor Schauberger noted how springs emerge under cover, never in full sunlight. In fact, he found that springs tended to recede when they were exposed to the Sun.[3] This is a good illustration of water's preference for darkness and indirect light. The banks of healthy rivers are always heavily vegetated, and when plant life is removed, the water and the entire ecosystem suffer.

The ocean is an ideal example of how sunlight is balanced with darkness to contribute to highly structured water. UV rays break hydrogen bonds in the upper 35 m (115 ft) of the ocean and deliver new information to water every day. IR rays from the Sun and Earth continuously reorganize the structured water matrix below. Although water from the ocean has many more minerals than freshwater, it is *mature* and therefore does not form scale like *juvenile* freshwater does.

Color frequencies in water

Water is the ideal medium for the transmission of many forms of information, including the full spectrum of colors. For hundreds of years color therapies, known collectively as hydrochromopathy, have used water's ability to transfer the information from color. Colors generate magnetic fields and electrical impulses that activate biochemical and hormonal processes in the human body.[4] American physician Edwin Babbitt (1828–1905) developed a comprehensive theory of healing using light and color. His book, *The Principles of Light and Colour*, prescribes different colors for the treatment of a variety of symptoms. Babbitt also developed *elixirs* by exposing water to sunlight filtered through colored lenses. He asserted that water "potentized" by specific light frequencies had healing properties.[5]

Have you ever noticed how magical the light is underneath the foliage in a forest? Have you noticed how the diffuse light in the shade of a tree contains the aspects of many colors? Sometimes as light filters through a canopy of leaves, you can see rainbows. This is because light is filtered by plants just like it is by water. Some of the Sun's light is absorbed and some is reflected. Each plant's color spectrum and DNA attract and filter light in its own unique way and then *feeds* the life around it with various hues and intensities of filtered and reflected light.

Light and DNA

Vladimir Poponin explored DNA's electromagnetic properties and unique ability to attract light. His work reveals that DNA weaves light around itself.[6] This may also explain why the light filtered through plants and organic materials can be so pleasing and therapeutic. DNA takes light on a spiraling journey through the colored lenses of leaves, flower petals, and other natural light carriers. What emerges is light that is naturally harmonious with life.

DNA and structured water

In many ways, the structures of DNA and liquid crystalline water are similar. Both contain molecular rings held together by hydrogen bonds and what is called *base stacking*. In the graphic to the right, observe how layers of hydrogen-bonded water rings can be stacked in a helical arrangement similar to that of DNA. This could be one of the reasons both structured water and DNA absorb light in the 260–270 nanometer (nm; one billionth of a meter) wavelength range.* Structural similarity results in similar absorption spectra.

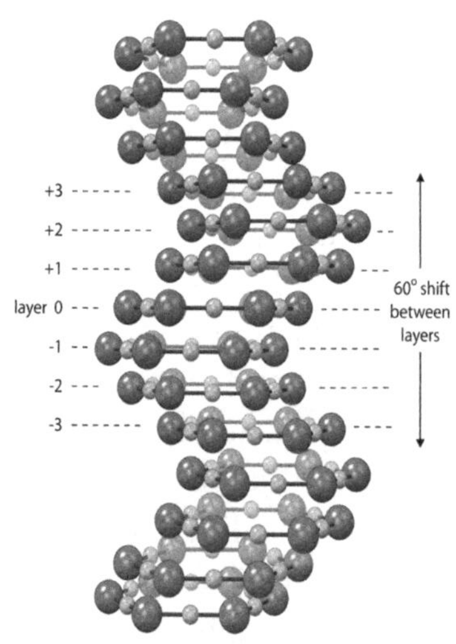

* Pollack found that when water formed highly structured zones, it uniquely absorbed light in a small portion of the UV spectral range (260–270 nm) (Pollack, G. *The Fourth Phase of Water*, pp. 34–45). Unstructured water did not show this peak. Nucleic acids (DNA and RNA) also absorb light in the 260–265 nm wavelength range (Niggli, H. Biophotons: Ultraweak light impulses regulate life processes in aging. *Gerontology and Geriatric Research* 2014 3:2).

The Sun and stars deliver valuable information carried on light vibrations to water on Earth. All this information is stored and renewed daily as water dances with light, sound, minerals, and other creative forces.

Part II provides details on using light to transfer information and to infuse color frequencies in water. The next chapter discusses the effects of sound on water's structure.

Image Credits (Chapter 7)

Page 84 Hydrogen-bonded water rings
Pollack, G. (2013). *The Fourth Phase of Water,* Ebner & Sons

References (Chapter 7)

1 Pollack, G. (2013). *The Fourth Phase of Water: Beyond Solid, Liquid and Vapor*, Ebner & Sons p. 87.

2 Lu, Z. (2006). Optical Absorption of Pure Water in the Blue and Ultraviolet. A dissertation submitted to the Office of Graduate Studies of Texas A&M University.

3 Bartholomew, A. (2005). *Hidden Nature: The Startling Insights of Viktor Schauberger.* Adventures Unlimited Press, p. 26.

4 Azeemi, S. and Raza, S. (2005). A Critical Analysis of Chromotherapy and Its Scientific Evolution. *Evidence-Based Complementary and Alternative Medicine*, 2(4), pp. 481–488. Available online: http://www.ncbi.nlm.nih.gov/pmc/articles/PMC1297510/

5 Azeemi, S. and Raza, S. (2005). A Critical Analysis of Chromotherapy and Its Scientific Evolution. *Evidence-Based Complementary and Alternative Medicine*, 2(4), pp. 481–488. Available online: http://www.ncbi.nlm.nih.gov/pmc/articles/PMC1297510/

6 Poponin, V. The DNA Phantom Effect: Direct Measurement of a New Field in the Vacuum Substructure. Available online: http://www.stealthskater.com/Documents/Consciousness_29.pdf

Chapter 8

The Influences of Sound and Vibration

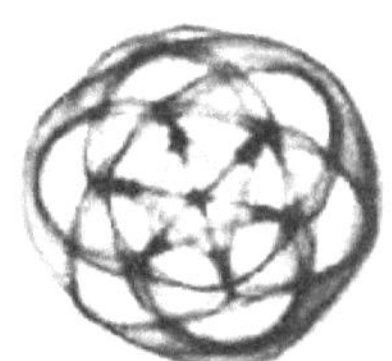

Rudolf Steiner

Sound and vibration affect water's structure to a greater degree than light does. According to Dr. Rudolf Steiner (1861–1925), Austrian philosopher and founder of biodynamics and anthroposophy, water is the ideal carrier for sound. Because the Earth is predominantly water, it is the perfect receptacle for the creative tones of the universe, including the sounds of the natural world and the resonance of the Earth. Although sound adds vibrancy and life, it can also provide the *energy* required to bring liquid crystalline structure to water.

Pollack and his team at the University of Washington observed that as soon as applied sound was removed, the liquid crystalline water network began to assemble. Frequencies in the experimental range led to the expansion of the ordered water network by five and six times its previous size.[1] They theorized that sound caused water molecules to disassociate—preparing the way for the development of a more organized matrix.

Sound and vibration as organizing forces

The Swiss natural scientist and physician Hans Jenny (1904–1972) demonstrated how sound and vibration create form and structure. His work, which he called *cymatics*, used audible frequencies to excite powders and liquids into lifelike images that varied from stationary patterns to those in continuous motion. He generated repeatable geometry with golden ratio symmetry.

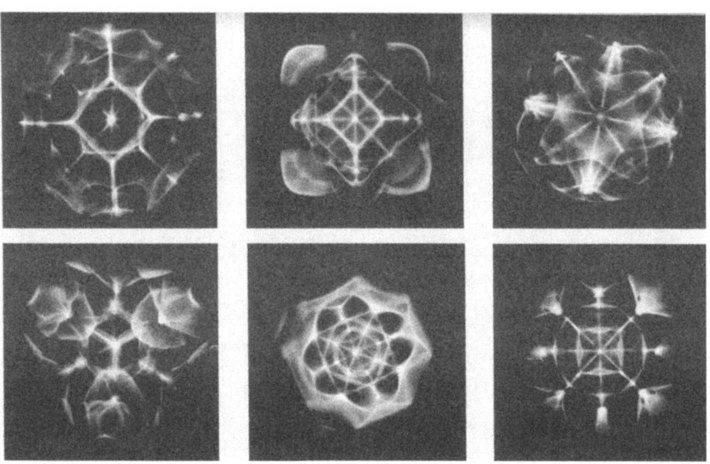

Standing Wave Patterns Generated in Water
From *Cymatics: A Study of Wave Phenomena and Vibration*, by Hans Jenny

Jenny's work illustrates how Nature uses chaos (breakdown) as a bridge to creation. With rising amplitude, or frequency, old patterns broke down, making way for new patterns and more complex levels of organization. Jenny concluded that "The creative power . . . is inherent in tone, in sound."[2] Mae-Wan Ho expressed a similar idea. She theorized that the *flow of energy* is the organizing

principle in nature that creates and sustains life.[3] The flow of energy is characterized by vibration, which is the *music* in the dance with water. A visual demonstration of how sound frequencies organize and reorganize matter is available on YouTube:
http://www.youtube.com/watch?v=Qf0t4qIVWF4&feature=related.

Nature's vibratory symphony

According to Steiner, "All of nature begins to whisper its secrets to us through its sounds." We are vibratory beings—the result of many different frequencies gathered to create organic life. Increasing levels of biological complexity are characterized by a greater number of frequencies.[4] You might wonder how different frequencies can combine without creating interference and chaos. The interesting thing is that when frequencies are added in golden ratio harmony, an infinite number can be assembled without cancellation or distortion.[5] Nature uses golden ratio harmonics to create complex and intricate structures. In other words, every natural creation is a vibratory symphony played in golden ratio harmony.

The connection between math, geometry, and frequency

All the frequencies in the electromagnetic spectrum are related in octaves. An octave is an interval with half or double the frequency of the original note. For example, if the original note has a frequency of 18 Hz, the same note an octave above it will resonate at 36 Hz, and the same note an octave below will resonate at 9 Hz. The term *Hertz* (Hz) means cycles per second. This is *not* an arbitrary designation. It is an increment of the time it takes the Earth to complete its daily rotation. Although Hertz is a modern term, the increment it represents is a function of our fractal universe.

Eric Rankin has demonstrated the connection between mathematics, geometry, and frequency. His work reveals how Platonic geometry, the Flower of Life, and the golden ratio proportion all play the same basic *chord*.[6] By adding the angles of basic shapes and converting the resulting number to a vibratory cycle (frequency in Hertz), you can "hear" the tone produced by each shape. Rankin discovered that an equilateral triangle (three

60-degree angles) produces a frequency of 180 Hz; a square (four 90-degree angles), 360 Hz, exactly one octave above the frequency of a triangle; a pentagon (five 108-degree angles), 540 Hz, a perfect harmonic fifth above the original note of the triangle; a hexagon (six 120-degree angles), 720 Hz, another octave above the original note; a heptagon (seven 128.57-degree angles), 900 Hz, a harmonic third; and an octahedron (eight 135-degree angles), 1080 Hz, another harmonic fifth.

Rankin discovered that the chord is repeated in the 3-D Platonic solids, where angles in the solids add to the same notes in different octaves.

Tetrahedron = 720 Hz
Cube or hexahedron = 2160 Hz
Octahedron = 1440 Hz
Icosahedron = 3600 Hz

Going further, Rankin took the 360-degree circumference of a circle and added it incrementally to create the seed of life (the building block of the Flower of Life): 360, 720, 1080, 1440, 1800, and 2160 notes to the same chord.

Then Rankin looked at the Fibonacci sequence (see page 72). When 180 is used as the initial number, the sequence proceeds 180, 180, 360, 540, 900, 1440, and so on, and results in the same chord. Any frequency from the chord used to begin the sequence always produces the identical three notes (in different octaves). The same chord emerges from the golden ratio spiral, where the *whole* increases by a factor of phi at each quarter turn (90 degrees): 90, 180, 270, 360, 450, 540, 630, 720, 810, 900, and so on.

The F# Major Chord Derived from Geometry

F# = 180 Hz, with octaves at 360, 720, 1440, 2880, etc.
A# = 225 Hz (harmonic 3rd), with octaves at 450, 900, 1800, 3600, 7200
C# = 260 Hz (harmonic 5th), with octaves at 540, 1080, 2160, 4320

One of Rankin's most revealing experiences occurred during his visit to the world-famous Integratron, an acoustically perfect sound chamber. Following a recording session, he played the note F# (equivalent to 740 Hz in the international standard tuning method, where A = 440 Hz). Letting the tone gradually descend, he noticed that when it reached 720 Hz, he felt as though everything came into sync. He described the experience as "like coming home."[7] Rankin realized that 720 Hz was the note F# derived from geometry. Interestingly, the F# major chord, when played in another tuning where A = 432 Hz, comes closest to matching the numeric values Rankin discovered in geometry. Rankin's work (found at sonicgeometry.com) invites deeper investigation into the relationship between math, geometry, and frequency.

432 versus 440 tuning

Although the world has settled on an international tuning standard that places the note A at 440 Hz, Rudolf Steiner and others advocated the slightly lower tuning: A = 432 Hz.[8] The difference is barely perceptible to the untrained ear; however, most people who listen to music played in both tunings say that 432 Hz tuning produces warmer, richer tones and music that is more pleasing. There are plenty of examples on the Internet where selections played in both tunings can be compared—and a great deal of evidence gathered by author Maria Renold to show that 432 Hz tuning consistently produces a more relaxed audience when compared to 440 tuning.[9] Her years of research with individual tones (played by a variety of instruments) reveal that over 90% of the individuals from four countries preferred the quality of individual tones associated with 432 Hz tuning.[10]

There are dozens of tuning methods and just as many ways of creating a musical scale. Maria Renold identified a method that combines 432 Hz tuning with the unique *scale of twelve fifths.* This scale offers an alternative to the false intervals created by equal-tempered tuning in modern music, yet it retains the advantages of a chromatic scale. Her method places the note C at 128 Hz, as advocated by Steiner, who said that the human ear was based on C = 128 Hz.[11] Doctors and ear specialists still use 128 Hz and 256 Hz (the higher octave) tuning forks for medical purposes, presumably because they resonate with the human body.

Tuning based on A = 432 Hz is a more natural tuning. It resonates with the organic universe, whereas 440 Hz has no harmonious connection to the natural world.[12] The authors of *Dancing with Water* have conducted simple experiments with a chromatic tuner that reveal natural materials and sounds from Nature (birds, crickets, the human voice) often deliver harmonics in the 432 Hz tuning spectrum. This supports the recommendation that natural materials and sounds from Nature be used preferentially in the dance with water. Digitally reproduced sounds (as good as they may be) lack the qualitative content of the natural, organic world.

Resonance in the natural world

Not only is vibration an element of the creative process; it is also a stabilizing force. The relaxed human body vibrates at a rate of approximately 8 Hz (a lower octave of 128 Hz). The Earth also vibrates at the same approximate frequency, known as the Schumann resonance. It is quite well known that the resonance of the Earth is balancing for all organic life forms. When we are in direct contact with the Earth, we are supported via resonance, but when we are disconnected from the Earth and surrounded by unnatural frequencies, our bodies have a difficult time staying in *tune*. Diseased organs always vibrate at unnatural frequencies, and if they are not *retuned*, they can set off a slow deterioration of the whole organism.

The natural world continuously *retunes* itself. When a bee pollinates a flower, it does much more than take nectar to make honey. The beat of its wings provides stimulatory information for the plant. And just as the beat of a bee's wings provides vibratory nourishment, the song of a bird is an integral part of the music composed by Mother Nature. Research illustrates the healing effects of the sounds in nature. They are fundamental to life.[13] Even sitting on a granite boulder on a sunny day provides a broad range of audible and inaudible frequencies—everything from the rustle of the trees in the wind to birds and insects and the frequencies of the minerals in the granite.

Vibratory nourishment

The interesting thing about resonance is that if several frequencies are provided, objects or organisms tend to recognize the frequency(ies) they need. For example, if you tap a tuning fork in a room full of tuning forks, only the ones with the same frequency will vibrate. Similarly, if a blend of frequencies is provided for the human body, each organ or system will select the frequencies it can benefit from. Harmonious vibratory information supplied on a consistent basis retunes and restores health. Researchers such as Royal Rife and Hulda Clark used frequencies to resolve many ailments. An extension of their methods today is the use of pulsed electromagnetic field (PEMF) devices, which provide a multitude of harmonic pulses.

The liquid crystalline structure of the human body (and all biological life) was designed to absorb and respond to vibratory information delivered through water. Vibration itself can be considered a nutrient. According to some, it is *vibration* that nourishes every cell and provides the basis for health.[14] Sometimes, just the frequency of a vitamin or mineral is enough. In her book *Return to Harmony*, Nicole LaVoie elaborates on the use of frequencies to nourish and balance the body. Her research demonstrates the correlation between the absence of certain frequencies in the voice and corresponding health difficulties. According to LaVoie, the missing vibratory *nutrients* can be supplied in water.[15]

A good way to stay in harmony with the natural world is through the water you drink. As a carrier of vibratory information, water can be instrumental in maintaining the organic symphony of life. Informed with vibratory input from the natural world, water becomes *full-spectrum living water.* Without it, water is incomplete.

Not only does sound play a role in the development of water's liquid crystalline structure; it can also be used to add information. Water can be enhanced using audible sound (music, toning, drumming, etc.) and with many inaudible vibrations from Nature. In Part II you will learn how to use sound and vibration to structure and add nutritive information to water.

Image Credits (Chapter 8)

Page 88 Standing Wave Patterns Generated in Water
Cymatics: A Study of Wave Phenomena and Vibration, by Hans Jenny. ©2001 MACROMedia Publishing, Eliot, ME
www.cymaticsource.com

References (Chapter 8)

1 Pollack, G. (2013). *The Fourth Phase of Water: Beyond Solid, Liquid Vapor*, Ebner & Sons, pp. 90-91.

2 Jenny, H. (2001). *Cymatics: A Study of Wave Phenomena and Vibration,* Newmarket, NH: MACROMedia. This edition is a compilation of the original two volumes published in 1967 and 1974.

3 Ho, M-W. (1998). *The Rainbow and the Worm*, World Scientific, pp. 40 and 75-77.

4 Roffe, R. Harmonic Resonance. Available online: http://www.rexresearch.com/articles/roffe.htm

5 Murphy, B. (2012). *The Grand Illusion: A Synthesis of Science & Spirituality*, Balboa Press, p. 179.

6 Rankin, E. http://sonicgeometry.com

7 Rankin, E. Personal conversation, March 20, 2015.

8 Renold, M. (2004). *Intervals, Scales, Tones and the Concert Pitch C = 128 Hz*. Temple Lodge Publishing English Translation, p. 161.

9 Renold, M. (2004). *Intervals, Scales, Tones and the Concert Pitch C = 128 Hz*. Temple Lodge Publishing English Translation, p. 69.

10 Renold, M. (2004). *Intervals, Scales, Tones and the Concert Pitch C = 128 Hz*. Temple Lodge Publishing English Translation, pp. 74-78.

11 Renold, M. (2004). *Intervals, Scales, Tones and the Concert Pitch C = 128 Hz*. Temple Lodge Publishing English Translation, pp 80 and 107-125.

12 Renold, M. (2004). *Intervals, Scales, Tones and the Concert Pitch C = 128 Hz*. Temple Lodge Publishing English Translation, p. 125.

13 Tompkins, P. and Bird, C. (1990). *Secrets of the Soil,* Harper and Row, pp. 128-145.

14 Pischinger, A. (2007). *The Extracellular Matrix and Ground Regulation,* North Atlantic Books.

15 LaVoie, N. (1996). *Return to Harmony,* Sound Wave Energy Press.

Chapter 9

The Influences of Magnetism and Electricity

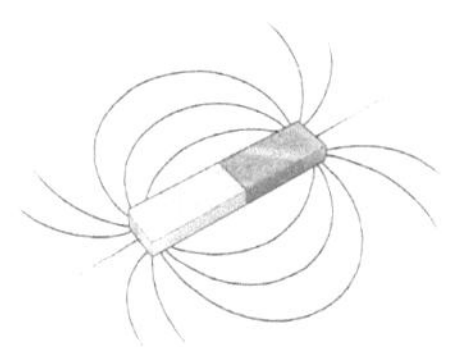

In the 1950s, Albert Szent-Györgyi postulated that water and electromagnetic fields form the matrix of life.[1] Today, there is no question that electric and magnetic fields influence the structure of water.[2] The forces of electricity and magnetism are different expressions of the electromagnetic "whole." Although they can never truly be separated, they are used as seemingly separate forces. As such, they have different effects on water.

Because water molecules are polar, they can be aligned in an electric or magnetic field. Many students are familiar with the "water bridge" formed between two separated containers of water as an electric current is passed between them. Electricity forces water molecules to become aligned within the electric current; however,

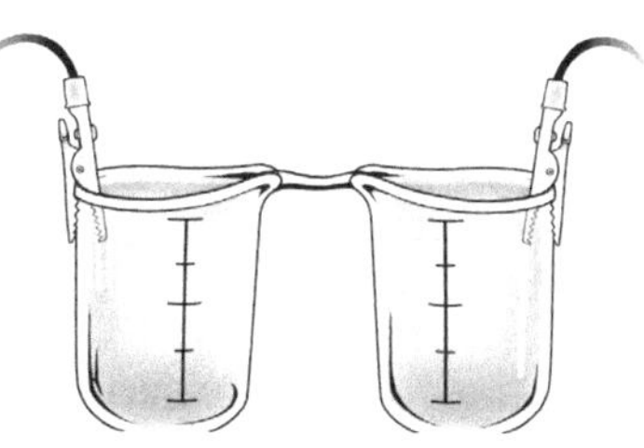

the alignment is *only in one direction*. It causes an overall reduction in hydrogen bonding.[3] The opposite occurs when water is exposed to a magnetic field. Magnetism increases hydrogen bonding and stabilizes the entire water network, even after the magnetic field has been removed.[4,5,6]

Some researchers originally proposed that electricity has a lasting effect on water's structure (as with the use of water ionizers).[7] This idea was based on the supposition that the smaller molecular units resulting from exposure to electricity were correlated with structured water and that these small molecular units were more easily absorbed by biological organisms.[8] Further research has revealed that small molecular units are associated with water that is poorly hydrogen bonded[9,10] and lacking in coherent organization. The ability of such water to carry signals and other energetic information is hindered because the matrix is not organized throughout. Structured water (the coherent liquid crystal) is characterized by increased hydrogen bonding and decreased molecular motion (stability), which create a flexible, responsive network similar in many ways to that found in a solid crystal.

Not all forms of electricity are detrimental to water's organization. The strength of the current can make a difference.[11] Dr. Pollack has identified zones of organized water with an abundance of electrons coexisting next to zones of water with an abundance of protons in biological tissues. Together, these are responsible for a persistent flow of current.[12] Another example of weak electrical forces with a positive effect on water's structure is the currents generated as water moves through cell membranes. These currents provide a source of energy for the human body.[13] The movement of water through cellular membranes also *requires* electrical forces. Water can be absorbed into cells only when an electric charge, known as *membrane electrical potential*, exists at the cell membrane. Water enters the cell one molecule at a time (not in small clusters) through a channel called an aquaporin.[14]

Piezoelectricity

Piezoelectricity is Nature's subtle form of electricity. It results when normally nonconductive materials are polarized by pressure (*piezo* = Greek for "pressure"). The pressure creates a conductive pathway and a gentle voltage. In the human body, piezoelectricity results when pressure waves polarize water and other liquid crystalline biomolecules.[15] The subtle voltage contributes to water's structure.

Quartz crystals also produce piezoelectric current when slight pressure is applied. As water flows over sand (mostly quartz and other crystalline minerals) gentle piezoelectric currents participate in the continuous structuring of the Earth's water. Quartz has been used for centuries to treat water. Its effects are subtle yet potent.

Alternating current

The weak piezoelectric currents generated within the human body and on the surface of the Earth are much different from the man-made electricity (alternating current) used to power our modern electrical grid. Alternating current (AC) induces voltage in nearby conductive objects, including biological organisms. Considerable evidence suggests this current may interfere with the subtle electric and piezoelectric currents within the human body and may disrupt biological processes.

Water molecules within several feet of electrical wiring or appliances carrying AC undergo rapid movement, reversing polarity 50 or 60 times/second, depending on whether you live in a country with 50-cycle or 60-cycle current. This rapid movement causes water to heat up. Microwaves are the worst because they reverse the polarity of water molecules 2½ billion times a second[16] and produce so much friction that water (and water in food) heats up extremely fast. Microwave ovens obliterate water's structure and zap its life force within seconds. No wonder Swiss, German, and Russian researchers determined that microwaved food is lacking in food value.[17]

Water should not be left unprotected in close proximity (within several feet) of electrical wiring, electrical appliances (especially microwaves), computers, Wi-Fi connections, cell and cordless phones, or fluorescent lighting—even when they are turned off. The electromagnetic frequencies that accompany these devices have a disruptive effect on water's structure and delicate electromagnetic field. Additionally, water is not easily absorbed (taken in at a cellular level) in the presence of these unnatural electromagnetic fields, which could be a factor in the chronic dehydration experienced by a growing percentage of the population. In our modern world, we are almost never in an environment that is free from electrical fields unless we are outside and away from power lines, cell phones, and satellite reception—or unless we have put measures in place to protect ourselves from these influences.

Direct current

Direct current (DC) is different. It is a one-way current flow. It does not affect conductive objects in the vicinity the way AC does, so it is considered to be more safe. However, strong DC voltages, like those used to electrolyze/ionize water, are also harmful to water's structure. Although they force molecular alignment in the direction of the current, they disrupt overall organization and break bonds that stabilize the water network. Originally, DC was thought to bring structure to water, but any degree of structure created with DC is directional; it is short-lived and not coherent (for more information on water ionizers see Appendix B: The Electrolysis of Water).

Water and magnetic fields

Magnets have been used to treat a variety of conditions in the human body for hundreds of years. They have also been used to structure water. Magnets generate spiraling vortices in moving water (and in the bloodstream) based on the interaction of magnetic forces with ions/salts (charged particles). However, even distilled and deionized water respond to magnetism because of water's inherent polarity.

When placed within a magnetic field, water molecules orient themselves in the most efficient manner to withstand the internal "pressure" created by the magnetic field. Unlike being immediately *trapped* in an *electric* field, water molecules within a *magnetic* field are able to orient themselves in an organized geometry that maximizes the whole water network. Magnetic fields encourage hydrogen bonding and the development of a coherent matrix as long as the magnetic field is not overpowering. The process is similar to the way carbon molecules become organized over time to withstand the pressure deep within the Earth as a diamond develops. Magnets can be used in several ways to bring coherent structure to water.

A variety of experiments have shown that both standing and moving water are affected by magnetic fields. Thus, simply placing a magnet beneath or alongside a glass of water will affect its structure in time. To maximize the effects of magnetism, water is most often passed through a magnetic field or cycled within a magnetic field. Cycling or vortexing within a magnetic field erases "information" in water. With continued cycling, molecules develop coherent order, much like a school of fish that instinctively returns to order after a disturbance.

Paramagnetism

Natural magnetic forces are divided into three categories: ferromagnetic, diamagnetic, and paramagnetic. Ferromagnetic substances (iron, nickel, cobalt, and some alloys) are strongly attracted to a magnet. They retain their atomic or molecular alignment after the magnetic field has been removed.

Diamagnetic materials have electrons that are all paired so that the opposite spins of electrons cancel each other out; these substances are not attracted to a magnet, but they are influenced by the presence of magnetic fields. Atoms in diamagnetic substances align *against* the direction of a magnetic field. Most organic compounds, including water, are diamagnetic. Water is actually repelled by magnets.

Paramagnetic substances are weakly attracted to a magnet. They align with the direction of a magnetic field, but they do not retain their alignment when the magnetic field is removed. Paramagnetic substances have unpaired electrons that may spin at a higher than normal velocity generating a permanent magnetic field around each atom, where each one becomes a weak magnet. When paramagnetic materials are placed within a magnetic field, molecular alignment occurs in the direction of magnetic force, and the overall magnetic field is amplified.

There are many paramagnetic materials—some are more paramagnetic than others. The degree to which a material is either magnetic or paramagnetic is determined by its magnetic susceptibility. Magnetic susceptibility can be measured in CGS units (centimetergram-seconds), where CGS = the number of grams of material that move 1 cm. toward a magnet in one second. Volcanic rock (especially basalt and granite) is highly paramagnetic. It has a magnetic susceptibility rating from 200 μCGS (millionths of a CGS) to about 9000 μCGS, depending on the source of the rock.

Dr. Philip Callahan did much of the original research on paramagnetism as it relates to healthy life on Earth. As a radio technician during WWII, Callahan became interested in the extremely low frequency (ELF) vibrations of the Earth and their relationship to organic life. Callahan found evidence that paramagnetic materials amplified ELF emissions. His measurements at the round towers in Ireland, which are made of paramagnetic rock, revealed emissions three to eight times higher than normal ELF levels.[18] Furthering his work, he noted similar ELF amplification at many other paramagnetic sites all over the world.

Callahan's years of research have added to our understanding of the function of the resonant cavity between the Earth and the ionosphere. It was no surprise to him that the ELF frequencies produced in this cavity (called Schumann resonances, which the authors of this book refer to as Earth resonance*) are the same frequencies that predominate in the resonant cavity of the human skull (human brain waves).[19] These ELF frequencies have been reported to generate the coherent, fully hydrogen-bonded water network.[20] In other words, the resonance of the Earth itself brings structure to water.[21]

It makes perfect sense that the resonance of the Earth would support the coherent liquid crystalline structure of her most prized possession—water. Paramagnetism is uniquely tied to the Earth's magnetic field and to the resonance of the Earth, but that is not all. According to Callahan, "the paramagnetic forces of rock amplify not only ELF radio waves in the atmosphere . . . but also the photon waves generated in the infrared and visible control region of life."[22] Thus, paramagnetic materials amplify two of the most important frequency bands for structuring water, ELF and infrared (IR). Finely ground paramagnetic rock is the equivalent of millions of weak magnets amplifying the greater magnetic field of the Earth. This material amplifies Earth resonance and IR wavelengths, which both bring structure to water.

Paramagnetism and ormus

Paramagnetic materials ionize many of the minerals in water so they can more easily be held within its structure (the water is brought to

---

* The resonance of the Earth (Schuman resonances) provides a background pulsation that has been referred to as the tuning fork for life. It is a heartbeat-like frequency that balances all life on the planet and naturally synchronizes many hormones, brain waves, and biological rhythms. When we are physically disconnected from the Earth, we experience a variety of side effects including anxiety, disorientation, and sleep disturbances. Early astronauts experienced these and other physiological symptoms as a result of their disconnection from the Earth. Research conducted by Cyril Smith reveals that the removal of water from the geomagnetic field of the Earth causes water to lose its stored information.

maturity). Paramagnetism also spins the ormus elements into their high-spin state,[23] which helps water hold its structure. According to Barry Carter, the ormus elements exhibit paramagnetic properties.[24] It is likely that paramagnetism is linked to the presence of ormus in rocks and soil, which explains why paramagnetic sand placed around water activates the ormus elements and stabilizes the coherent liquid crystalline matrix.

Invertebrates, including sponges, sea spiders, starfish, and amphipods, grow many times their usual size in Antarctic water. Amphipod crustaceans in the Antarctic Ocean are more than five times as long as the largest temperate species.[25] Perhaps this is because water at the polar regions of the Earth is subjected to a magnetic field that is nearly double the strength of the magnetic field at the Earth's equator.[26] Plant growth is positively influenced by magnetic and paramagnetic fields,[27] and many human symptoms respond to magnetic field treatment. Invertebrate size and positive plant growth may be due to the effects of magnetism on the structure of water and on water's ability to carry the resonance of the Earth.

In Part II you will learn how to use magnetic, paramagnetic, and piezoelectric forces to make *full-spectrum living water* and how to protect your water from man-made electromagnetic fields.

References (Chapter 9)

1 Szent-Györgyi, A. (1957). *Bioenergetics*. Academic Press, p. 139.
2 Chaplin, M. LSBU website: http://www1.lsbu.ac.uk/water/magnetic_electric_effects.html
3 Vegiri, A. (2004). Reorientational Relaxation and Rotational-translational Coupling in Water Clusters in a DC External Electric Field. *Journal of Molecular Liquids*, 110(1) pp. 155–168.
4 Wang, Y. et al. (2013). The Effect of a Static Magnetic Field on the Hydrogen Bonding in Water Using Frictional Experiments, *Journal of Molecular Structure*. 11(1052), pp. 102-104.

5 Cai, R. Yang, H. He, J. and Zhu, W. (2009). The Effects of Magnetic Fields on Water Molecular Hydrogen Bonds, *Journal of Molecular Structure*, 938(1), pp. 15–19.
Available online: http://www.sciencedirect.com/science/article/pii/S0022286009005559

6 Chang, K-T. and Weng, C-I. (2006). The Effect of an External Magnetic Field on the Structure of Liquid Water Using Molecular Dynamics Simulation. *Journal of Applied Physics* 100(4). Available online: http://www.fluxindia.com/images/magneticinductor/references/effect.pdf

7 Hayashi, H. (1996). *Microwater, the Natural Solution*; Water Institute: Tokyo, Japan.

8 Hayashi, H. Understanding Alkaline "Ionized" Water. Online http://www.heartspring.net/water_science.html

9 Zhao, L. et al. (2015). Changes of Water Hydrogen Bond Network with Different Externalities. *International Journal of Molecular Science*, 16(4), pp. 8454-8489.

10 Chang, K-T. and Weng, C-I. (2006). The Effect of an External Magnetic Field on the Structure of Liquid Water Using Molecular Dynamics Simulation. *Journal of Applied Physics* 100(4). Available online: http://www.fluxindia.com/images/magneticinductor/references/effect.pdf

11 Chaplin, M. LSBU website: http://www1.lsbu.ac.uk/water/magnetic_electric_effects.html

12 Pollack, G. Figueroa, X. and Zhao, Q. (2009). Molecules, Water, and Radiant Energy: New Clues for the Origin of Life. *International Journal of Molecular Sciences*, 10(4), pp. 1419-1429.

13 Batmanghelidj, F. (2003). *Water for Health, for Healing, for Life*, Warner Books, pp. 36-37.

14 Agree, P. (2006). The Aquaporin Water Channels, *Proceedings of the American Thoracic Society*, 3(1), pp. 5–13. Available online: http://www.ncbi.nlm.nih.gov/pmc/articles/PMC2658677/

15 Oschman, J. (2003). *Energy Medicine in Therapeutics and Human Performance*, Butterworth-Heinemann, pp. 152-153 and 270.

16 Hyper Textbook Website: http://hypertextbook.com/facts/1998/HowardCheung.shtml

17 Georgiou, G. The Hidden Hazards of Microwave Cooking. *Journal for the American Association of Integrative Medicine- Online:* http://www.aaimedicine.com/jaaim/apr06/hazards.php?printable

18 Callahan, P. (1995). *Paramagnetism—Rediscovering Nature's Secret Force of Growth*, Acres USA, p. 70.

19 Murugan, N. Karbowski, L. and Persinger, M. (2014). Serial pH Increments (~20 to 40 Milliseconds) in Water During Exposures to Weak, Physiologically Patterned Magnetic Fields: Implications for Consciousness. *Water Journal* 6(1), 45-60. Available online: http://waterjournal.org/uploads/vol6/persinger/WATER.2014.2.Persinger.pdf
20 Chaplin, M. LSBU Website: http://www1.lsbu.ac.uk/water/magnetic_electric_effects.html
21 Shen, X. (2011). Increased Dielectric Constant in the Water Treated by Extremely Low Frequency Electromagnetic Field and Its Possible Biological Implication. *Journal of Physics: Conference Series*, 329(1). Abstract online: http://adsabs.harvard.edu/abs/2011JPhCS.329a2019S
22 Callahan, P. (1995). *Paramagnetism—Rediscovering Nature's Secret Force of Growth*, Acres USA, p. 88.
23 Carter, B. Ormus and Paramagnetic Soils. Available online: http://www.subtleenergies.com/ormus/tw/paramag.htm
24 Carter, B Ormus and Paramagnetic Soils. Available online: mhttp://www.subtleenergies.com/ormus/tw/paramag.htm
25 Census of Antarctic Marine Life. Available online: http://www.caml.aq/voyages/tangaroa-2007-2008/logbook-week3.html
26 Hyperphysics Website: http://hyperphysics.phy-astr.gsu.edu/Hbase/magnetic/MagEarth.html#c1
27 Callahan, P. (1995). *Paramagnetism—Rediscovering Nature's Secret Force of Growth*, Acres USA, pp. 103-109.

It makes perfect sense that the resonance of the Earth would support the coherent, liquid crystalline structure of her most prized possession—Water.

Paramagnetic materials amplify two of the most important frequency bands for the creation of full-spectrum living water, the resonances of the Earth and the infrared frequencies from the Sun.

Chapter 10

Stillness
Nature's Meditation

Many natural forces (light, sound, geometry, magnetism, piezoelectricity) provide the energy for the development of water's structure. Some forces are responsible for freeing water molecules from old patterns and random hydrogen bonding; others contribute to the assembly of liquid crystalline order. The final step in the creation of *full-spectrum living water* is to allow water to remain still—in a cool, dark place exposed to the resonance of the Earth.

Breakdown followed by stillness

Breakdown is a necessary step in the structuring process. Masculine forces (turbulence, sunlight, heat, sound, and many electromagnetic fields) break hydrogen bonding. These forces loosen intermolecular connections.[1] They facilitate the release of old patterns and pave the way for coherence in the restructured medium. Then, during stillness, feminine elements (darkness, cool temperatures, and Earth resonance) help to build a coherent structured matrix to house new information. As the liquid crystalline matrix develops, surface tension and viscosity *increase.*[2,3] This is what gives structured water its smooth, creamy texture.

Think about how water flows in a river. Rocky areas, rapids, and waterfalls are often followed by deep pools where the water is allowed to sit in stillness. Wind works in a similar manner on land. It facilitates the release of stagnant energy and is always followed by stillness. One of the authors recalls the abatement of a windstorm one afternoon. After hours of violent turbulence, everything became still. The natural energy field around each tree began to expand. The author distinctly remembers feeling that the trees were in a state of meditation, integrating new information from their environment. The energy was almost palpable as the field around each tree grew to meet the energy field of its neighbor.

Wind and water work together for the plant kingdom. Storms cause violent movement that forces the water in each plant cell to release old patterns. Then, in stillness, updated information is assimilated from the environment for the next season or the next phase of development. In Nature, this cycle repeats over and over. It is a very similar process for bodies of water. After a preparatory phase of movement and release, water molecules find their place in a liquid crystalline array during stillness. This is the best time for water to incorporate new information.

It is interesting that an energy *field* develops around water too. The growing energy field supports and stabilizes the molecular structure of the water; at the same time, the developing structure of the water supports the growth and maintenance of the energy field. This is the reason, *full-spectrum living water* is able to hold its structure for a long period of time. The presence of an energy field around maturing water can often be just as palpable as the field that develops around trees after a storm. Water that reaches maturity in stillness resonates with life force and can influence the energy of an entire room.

Water that is allowed to remain still following a phase of movement will develop some level of maturity and coherence no matter where or how long it sits. However, there are ways to enhance the process so coherence is maximized and the resulting water is filled with life force. The key is to bring in the feminine elements: darkness, cool temperatures, and Earth resonance.

Darkness

Energy is required for both steps of the structuring process (breakdown and reorganization). Light often contributes to the structuring process (providing energy for breakdown) but darkness completes the equation. It is the infrared (IR) light spectrum, available in total darkness, that is the most potent wave band for inducing structural organization in water. Darkness and the energetic qualities of the womb *refine* and *mature* water's energy. Schauberger called this refining process "ennoblement." In a patent for the production of springwater-quality drinking water, Schauberger wrote: "It is also necessary that this process should take place entirely in the dark (insulated from light), since experiments have shown that similar processes of ennoblement under the influence of light yield inferior water."

The gestational period in many higher life forms is spent in the protective darkness of a womb. There, it is sheltered and nurtured in darkness until it has gathered enough life force to spring forth into the light. Even after life emerges, it must find regular rest and renewal in darkness. Darkness is regenerative for water just as it is for the human body and for all life. Consider the significance of the underground portion of the Earth's water cycle. Only in darkness within the Earth does water become fully mature before rising to the surface. The vast majority of the ocean also lies in complete darkness, where water's life force is cultivated. Dew is another example of mature, energized water created in darkness. It is enlivened by the indirect light of dawn, and yet, if not absorbed, its energy is robbed by the Sun.

Cool temperatures

Darkness is usually accompanied by cooler temperatures. Schauberger emphasized the importance of temperature during the ennoblement of water. A cool, dark environment supports the development of coherence—especially when the temperature approaches 4° C (39.2° F).[4] Under these circumstances, water reaches perfect balance and equilibrium as hydrogen bonding increases. At cool temperatures, gases are quiet and less aggressive, and ormus elements become stabilized within molecular water cages.

This does not mean that water must be refrigerated. Often, refrigeration places water in an electromagnetic environment that can be counterproductive. Additionally, many people do not like to drink cold water. The best way to keep water cool is to allow it to breathe. When a container is porous, evaporative cooling naturally occurs at the outer edges of the vessel. This cooler and denser water sinks to the bottom, forcing warmer water to rise up the center. The process enhances circulation and ensures constant cooling. Earthenware (ceramic) vessels are ideal for helping to keep water cool. However, as long as the temperature is not excessive and as long as it is not exposed to direct sunlight, water usually stays cooler than the temperature of the surrounding air. This environment supports the development of mature water.

Earth resonance

Just below audible sound, the Earth resonates at a base frequency of 7.8 Hz, with many harmonics. This frequency band influences all life and is extremely supportive of water. Placing an appropriate vessel directly on the Earth connects water with this nurturing vibration and assists the process of maturation. Unfortunately, it is not often realistic to place a water container directly on the Earth. Alternatively, the use of the Earth's minerals provides this resonance and its harmonics, effectively "grounding" water during stillness. Paramagnetic materials amplify the Earth's resonance (see page 101). Earthenware (clay) and glass (silicon dioxide/quartz) vessels also bring in the Earth's resonance. Other methods for bringing

Earth resonance to water include orgonite (see page 155), Earthing (see page 171), and the use of beneficial microbes.

Beneficial microbes

In the 1980s and 1990s a Japanese professor, Teruo Higa, discovered a group of microorganisms that balance the natural flora in soil and improve plant yields many times over fertilizers and other additives. (Microbes exist in symbiotic relationship with plant roots and participate in the production of fulvic acids, which in turn make nutrients available for plant use.) Subsequent research revealed the organisms could survive severe conditions and extreme temperatures.[5] This made it possible to incorporate them in ceramics, thus creating a new way to store and transport beneficial microbes. Dr. Higa named the microbes *effective microorganisms.*

Microbes referred to as extremophiles live deep within the Earth. They represent some of the oldest organisms on the planet. During volcanic eruptions they are brought to the surface, and after cooling, they can remain dormant within rocks for millions of years. When rocks break down, the surrounding soil is populated with beneficial organisms. Water that emerges from a pristine spring has some of the same microbes—supplied in minute quantities by igneous rocks as the water passes over them on the way to the surface. These microbes balance the flora in water and hold the resonance of the Earth.

Since Dr. Higa's initial discovery, other microbial populations have been found in volcanic ores. One product, referred to as QELBY, uses an ancient microbial community embedded in ceramic for amending soil and water. The addition of QELBY ceramics adds beneficial microorganisms to water. It also provides infrared wavelengths to help bring natural structure and draws hydrogen (as protons), and the DNA in the microbes attracts photonic energy. QELBY ceramics are a wonderful amendment where water will remain still for an extended period of time (overnight or longer) and a good way to add Earth resonance.

An egg-shaped container

Another important component of Schauberger's ennoblement process was the placement of water in an egg-shaped container. Nature typically provides a protected and confined space for gestation, a place where feminine energy is concentrated. The shape of the egg is prominent in seeds and grains. The womb is egg shaped, and even the human energy field is shaped like an egg. The shape of the egg is an ideal enclosure for water because it concentrates and preserves life force.

Many cultures (ancient and modern) have used egg-shaped vessels for storing grain and liquids. Similar vessels are used for culturing fermented foods. In some areas of the world, egg-shaped earthenware vessels are still used for water. Although they are fragile and present difficulties for transportation and storage, egg-shaped clay vessels are still recognized as superior for water storage. A properly oriented egg-shaped vessel is the perfect container for water to sit in quiet meditation. "Tuned" to the harmonics of reception and creativity, the shape of the egg helps water reach refinement in stillness.

Having been through a preparatory phase of release and then allowed to be still within a conducive environment, water develops coherent, liquid crystalline structure. In this state, water is extremely responsive (programmable) and able to hold an almost unlimited amount of energy and information. This is *full-spectrum living water*, ideal for personal consumption or agricultural use and for a wide array of other uses.

In Part II you will learn how to gather the elements that culminate in stillness for the creation of *full-spectrum living water*.

References (Chapter 10)

1 Pollack, G. (2013). *The Fourth Phase of Water: Beyond Solid, Liquid, Vapor*, Ebner and Sons, p. 91.

2 Chaplin, M. LSBU website: http://www1.lsbu.ac.uk/

3 Pollack, G. (2013). *The Fourth Phase of Water: Beyond Solid, Liquid, Vapor*, Ebner and Sons, p. 38.

4 Chaplin, M. LSBU website: http://www1.lsbu.ac.uk/

5 Higa, T. (1998). *An Earth Saving Revolution* translated by Anja Kanal, Sunmark Publishing.

Part II

Let the Dance Begin

Introduction to Part II

Now that you understand your dance partner, now that you are familiar with ways to support her movement and with ways to help her maintain balance, now that you know how to enhance her energetic qualities, you are in a position to dance.

Water requires a clean and open arena—an uncontaminated place to dance. Then the establishment of organization/structure provides a medium into which vibrational music can be absorbed.

Dancing with water is an exercise in creativity and discovery. There are no strict rules. There is only *the dance* when you let your imagination go.

Chapter 11

Selecting a Source of Water
Filtration and Purification

Just as a work of art requires a clean canvas on which to paint, water requires freedom from contaminants that would distort the expression of its full potential. Many options are available when considering a source of water, everything from bottled water to filtering or purifying your own. Each has its benefits and limitations.

It is well known that a large portion of the bottled water available today is only filtered or purified tap water and that regulations for the bottled water industry are less stringent than those for municipal water. On the other hand, there are many wonderful bottled springwaters. Although the authors of this book do not want to discourage the use of these rich water resources, they cannot consciously encourage their use either. In the long run, bottled water contributes to a large carbon footprint as it is bottled and transported around the globe. The water bottling industry also sometimes abuses precious water resources. Natural springwater that is personally harvested from a clean spring is ideal. Unfortunately, this kind of water is unavailable to most people. The authors encourage people to find a source of locally available water or a way to treat the water that comes from their tap. This chapter is a brief discussion of the options. It is not intended to be all inclusive.

Basic categories of filtration and purification

1. Adsorption

Adsorption refers to the ability of a filtering media to draw and hold contaminants.

Carbon is the most common adsorbant. As a filtration medium, it can manage a wide variety of contaminants. *Activated* carbon is a form of processed carbon with more surface area and greater efficiency and is the preferred form of carbon. Activated carbon is good for the removal of chlorine, chlorine by-products and organic compounds such as pesticides.

Activated alumina and bone char also come under the category of adsorbants. They are used specifically to remove fluoride.

Activated alumina is made from aluminum hydroxide that has been heated to produce a greater surface area. It is the most common fluoride removal medium and also removes arsenic. Activated alumina should be used within strict guidelines that include a reduced flow rate or long enough contact time for the above contaminants to be adsorbed.

Bone char is carbonized animal bone. When used as a filtration medium, the fluoride in water is attracted to the calcium in the bone char. Fluoride has a strong affinity for calcium—it is the reason fluoride causes dental and skeletal fluorosis. This type of medium must also be used within strict guidelines to allow water enough contact time to remove fluoride.

Unless activated alumina or bone char is used in a system that allows slow percolation, its usefulness for fluoride removal can be limited. When filters are new, these adsorbants perform well, but manufacturer claims are often overstated and point-of-use filters should typically be changed more often than recommendations suggest.[1]

2. Catalytic carbon

Catalytic carbon is another form of carbon filtration specifically designed to remove chloramines. The process *decomposes* rather than adsorbs chloramines. In the case of catalytic carbon, the surface is modified to speed up a reaction that eventually reduces chloramines to harmless chloride ions. Many catalytic carbon filters are a part of tank systems intended to filter the incoming water for a whole home.

3. Distillation

Distillation is an effective way to remove almost all impurities from water. When combined with carbon filtration, it will remove almost 100% of bacteria, viruses, cysts, and parasites. Distillation also removes heavy metals, minerals, radioactive compounds, and organic contaminants. The resulting water is pure, but it is also empty and requires remineralization and reconditioning to be returned to its full potential (see Chapter 13 on remineralizing and reconditioning purified water).

4. Ion exchange

Ion exchange is the process where one ion is traded or switched for a less harmful ion. Water softeners are based on this principle, where sodium replaces calcium and magnesium. Ion exchange is the most effective method to remove nitrates. It is also the way zeolite works to remove a wide variety of contaminants.

Zeolite is often referred to as a molecular sieve. It is a group of natural and synthetic aluminum silicates characterized by cage-like cavities. The framework of zeolite has a negative charge, balanced by positively charged ions inside. Zeolite releases calcium and sodium in exchange for heavy metals, pharmaceuticals, and other positively charged contaminants, including some radioactive compounds. Many filters now include a layer of zeolite.

5. Oxidation

Oxidizing agents work by stripping electrons and either destabilizing or neutralizing contaminants. They are extremely effective for killing harmful microbes and rendering many other contaminants biologically inactive. The oxidants suited for water purification include ozone, chlorine dioxide, and a biotite mineral extract sold under a variety of names. These agents can be very effective for purifying water when used in combination with carbon filtration to remove the inactivated contaminants.

Ozone (O_3) is a strong oxidant with a short life span. As Nature's best disinfectant, ozone is 50% stronger than chlorine and over 3000 times faster.[2] When ozone comes in contact with contaminants in water, the third oxygen atom is released, killing pathogens and neutralizing most contaminants. It forms oxygen radicals as it decomposes, so the use of ozone has secondary oxidizing potential. Oxygen (O_2) is the only by-product. Small home units are relatively inexpensive and easy to use with no harmful by-products. **Note:** Ozone will neutralize just about every contaminant in water, including heavy metals and fluoride compounds, but it will not eliminate nitrates.

Chlorine dioxide is another oxidant that has been used for water treatment for over 60 years.[3] It is approved for both the pretreatment and final disinfection of municipal water. Although the chemical name implies chlorine, there is no free chlorine in chlorine dioxide, so it does not have any of the negative effects of chlorine. Chlorine dioxide contains oxygen in an extremely active form that kills viruses, bacteria, and protozoa. The only by-products are chlorite and chlorate ions, which can be removed with activated carbon filtration.[4] Chlorine dioxide is ideal for small water supplies. In addition to being an effective disinfectant, it promotes coagulation of some contaminants, including fluoride, heavy metals, iron, and manganese, which can then be removed with a carbon filter. Many campers and hikers use tablets that produce chlorine dioxide for water purification because it is effective over a wide pH range with no harmful effects on human health. It has also been used as a component of mouthwash to control the bacteria that cause bad breath.[5,6]

Biotite mineral extract is sold under the names Adya Clarity, Biotite Concentrate, Drops of Balance, and Roxtract. This compound is a patented, concentrated ionic mineral solution extracted from biotite mica with sulfuric acid. It works as an oxidant and a coagulating agent, neutralizing contaminants and gathering them (via coagulation, much like chlorine dioxide) for removal by filtration. Depending on how these solutions are used, they can remove heavy metals and fluoride from water. The product called Adya Clarity has been slandered in an attempt to remove it from the market, but it is a valuable adjunct to the water purification scenario. An article on the *Dancing with Water* website offers a detailed explanation of how it works and how to maximize its effectiveness to purify small batches of water.

6. Reverse osmosis (RO)

Like distillation, a good RO system can provide pure water when used in combination with carbon filtration. Although RO is an option for heavily contaminated water or water that contains heavy metals, nitrates, or fluoride, RO is a very harsh water treatment. It is similar to the homogenization process for milk. During homogenization, fat molecules are broken into small fragments as milk is forced through a sieve-like apparatus. Disassembled fat molecules cannot re-aggregate. Similarly, RO forces water through a membrane designed to allow individual water molecules to pass. During the process, water's hydrogen bonding is destroyed as water molecules are torn from each other and minerals are stripped away. Like the fat molecules in milk, it is very difficult for water molecules that have been separated in this manner to re-aggregate and join again in a cooperative manner. RO water requires remineralization and a period of time for recovery before it is able to carry the finely tuned information necessary to support organic life. If you use RO water, make sure to read Chapter 13 on remineralizing and reconditioning purified water.

7. UV light

Certain wavelengths in the UV (ultraviolet) light band are known to inactivate DNA. When these wavelengths are supplied during water treatment, they inactivate the DNA of bacteria, mold, algae, viruses, and parasites so they cannot reproduce. In some cases, organisms are killed, but in other instances they are only rendered unable to reproduce—each microbe is different. Contact time, intensity, and wavelength are factors in the process. UV systems are not foolproof, and occasionally organisms regain the ability to reproduce.[7] UV systems should always be used in conjunction with filtration to physically remove problematic organisms.

Major issues with municipal tap water

If you decide to use your tap water, the first thing to do is to find out what is in it. Contact your local water treatment authority and get a report. The major concerns with tap water are the chemicals purposefully added (treatment chemicals and fluoride) and pharmaceuticals, for which there are no detection/removal protocols in place.

Treatment chemicals are first on the list of contaminants to be removed from tap water. Although some communities are more progressive than others (using ozone rather than chlorine or chloramines to disinfect water), a small amount of chlorine or chloramines is always added as water leaves the treatment facility to protect from contamination during transit.

Chlorine and its by-products, trihalomethanes (THMs), are easy to remove using carbon filtration. If chlorine is your only problem and you cannot filter all the water that enters your home, a point-of-use filtration device for your drinking water and an additional filter for your shower will suffice. If you prefer to bathe, consider the use of products that hang from the bath faucet so that water runs through them on its way into the tub. These products are designed to remove a significant amount of the chlorine and THMs from bath water, Jacuzzis, and spas. Activated carbon, RO, distillation, and oxidizing agents will treat or remove chlorine and THMs.

Chloramines are a combination of chlorine and ammonia. They are much more difficult to remove—and unfortunately they are becoming more common in municipal water. Chloramines erode rubber gaskets used to connect water fixtures, such as those in washing machines. They react with lead and lead solder in older plumbing, causing toxic levels of lead to be released in water. Although typical carbon filters will remove chloramines, the water requires a longer contact time. Catalytic carbon filters, RO, and distillation all remove chloramines. Ozone also eliminates chloramines. Vitamin C (25 mg per gallon) will neutralize chloramines prior to carbon filtration in small batches of drinking water and even in the bath (1000 mg in a medium tub).[8] Zeolite is suggested for the removal of residual ammonia when carbon filtration is used to remove chloramines.

Pharmaceuticals and over-the-counter drugs have become a widespread problem in municipal water today. Water treatment facilities are not set up to detect or remove them. In large communities where water is recycled, you must assume these contaminants are present. Zeolite, RO, distillation, and oxidizing agents remove and/or neutralize pharmaceuticals.

Fluoride is a difficult problem because the fluoride ion is small. Most filtration does not remove it. Fluoride is most commonly removed using activated alumina or bone char—discussed earlier in the chapter. A combination of RO and deionization removes nearly 100%. Distillation also removes fluoride. Oxidizing agents will reduce fluoride when applied properly and allowed to remain in the water for several hours. When followed by carbon filtration, oxidants can remove a significant amount of fluoride.

Major issues with private water sources

Individuals whose water comes from a private well often think they are *safe*. Their water does not typically have treatment chemicals, pharmaceuticals, or fluoride. Yet the water from underground aquifers has been compromised in many parts of the world—especially in agricultural and/or industrial areas. Much of the underground water on the planet is connected, just like the circulatory system in our bodies. Sometimes pollutants that enter the ground miles away are consumed by unsuspecting private well owners who live in an otherwise protected area. There is no monitoring of underground aquifers for private wells. The biggest concerns for private water sources are harmful microbes, heavy metals, pesticides, and nitrates.

Microbial and parasitic contaminants are becoming more common in private sources of water. Warm climates and summer months pose even more problems. Ultraviolet (UV) light is the most common method for treating bacteria, mold, algae, viruses, and parasites. Ozone and other oxidants also kill microbes, but these methods do not remove the contaminants that should be filtered out following treatment. Filters with a pore size of 1 micron or smaller (look for "absolute 1 micron filters") will remove harmful microbes, as will RO and distillation.

Heavy metals (arsenic, cadmium, lead, and mercury) can all be removed by zeolite, RO and distillation. Activated carbon removes mercury and reduces arsenic, cadmium, and lead. Oxidizing agents oxidize heavy metals for removal by filtration.

Pesticides (and other synthetic organic chemicals) fall into the class of contaminants known as *organic contaminants*, which are easily removed by carbon filtration. Most RO and distillation systems use carbon filtration to remove pesticide residues *before* treatment to preserve the life of filtration media and RO membranes. Ozone and other oxidants also oxidize most pesticides for removal by filtration.

Nitrates end up in ground water in agricultural areas where there is heavy fertilization and/or use of pesticides. They can also be released as industrial waste. Because the nitrate ion is stable and highly soluble it is not easily removed by filtration. Ion exchange, RO, and distillation are the best methods for removal.

Solving the contamination problem

Water filtration/purification systems are becoming more complex in an effort to accommodate the growing list of concerns. In the end, they are just Band-Aids. The real solution to our global water pollution problem will require thoughtful action and a change in our thinking. In the short term, we are obliged to find ways to filter and purify our water; in the long run we must be a part of the solution. Receiving pure, clean water the way Nature intended is the ultimate goal.

Each individual can make changes that will solve rather than augment the water pollution problem. Get in the habit of looking for ways to clean up the water in your area. Ask yourself these questions: Do I use toxic cleaning agents that end up down the drain? Do I use lawn chemicals and copious amounts of water to have a green lawn? Do I flush bags of salt through a soft water conditioning system? Am I in a position to influence the use of chemicals in agriculture or industry? Where/how could I be a positive influence? Each individual's contemplation of the above questions will support the long-term process of renewing the water on the planet and enhance your personal dance with water.

The chapters in the remaining part of the book assume that you are working with a contaminant-free source of water.

References (Chapter 11)

1 Michaud, C. and Slovak, R. (2009). Report released at the 2009 Annual Water Quality Association convention.

2 Eagleton, J. (1999). Ozone in Water Treatment: A Brief Overview. Available online: http://www.delozone.com/files/ozone-overview-drinkingh2o-1999.pdf

3 Augenstein, H. (1974). Use of Chlorine Dioxide to Disinfect Water Supplies. *Journal of the American Water Works Association,* 66(12), pp. 716-717.

4 Collivignarelli, C., Sorlini, S., and Belluati, M. (2006). Chlorite Removal with GAC. *Journal of the American Water Works Association,* 98(12) pp. 74-81.

5 Soares, L., Guaitolini, R., Weyne, S., Falabella, M., Tinoco, E., and da Silva, D. (2013). The Effect of a Mouthrinse Containing Chlorine Dioxide in the Clinical Reduction of Volatile Sulfur Compounds. *General Dentistry,* pp. 46-49. Available online: http://www.agd.org/media/156825/GenDent_Jul13_Soares.pdf

6 Lynch, E., Sheerin, A., Claxson, A., Atherton, M., Rhodes, C., Silwood, C., Naughton, D., and Grootveld, M. (1997). Multicomponent Spectroscopic Investigations of Salivary Antioxidant Consumption by an Oral Rinse Preparation Containing the Stable Free Radical Species Chlorine Dioxide (ClO_2). *Free Radical Research,* 26(3), pp. 209-234.

7 Mofidi, A., Rochelle, P., Chou, C., and Metah, H. (2002). Bacterial Survival After Ultraviolet Light Disinfection: Resistance, Regrowth and Repair. American Water Works Association Annual Conference and Exhibition. Available online:

http://www.researchgate.net/publication/244478252_Bacterial_Survival_After_Ultraviolet_Light_Disinfection_Resistance_Regrowth_and_Repair

8 San Francisco Water Power Sewer Website: http://www.sfwater.org/modules/showdocument.aspx?documentid=4125

The real solution to our global water pollution problem will require thoughtful action and a change in our thinking. In the short term, we are obliged to find ways to filter and purify our water; in the long run we must be a part of the solution.

Chapter 12

How to Structure Your Water

The new sciences reveal that the presence of *all the right elements* is not enough to create life. Only when the right elements are organized—with the capacity to conduct and store energy—can life emerge.[1] Water is at the heart of the process. Not only is water a living part of the natural world; it is the basis for all self-organization. When water's liquid crystalline organization is restored and/or maintained, water has a greater capacity to support the physical life forms on the Earth.

Perhaps you have noticed that the elements Nature uses to create water's liquid crystalline state gather and/or sustain life force. Consider the following examples:

1. Movement creates vortices, which naturally gather energy.
2. Minerals anchor the life force in water. For many years, water without minerals has been referred to as "dead."
3. Light, especially the infrared wavelengths, support life on this planet. Infrared wavelengths structure water in complete darkness.

4. The sounds of the Earth (Earth resonance) bring structure to water. These extremely low frequencies have been referred to as a *tuning fork for life*.
5. Many shapes gather life force, as research has illustrated. They serve as wave guides to focus and cohere the energy inherent in the universe and to bring it into the physical world.
6. Magnetism (especially paramagnetism) and gentle piezoelectric currents are also intrinsically linked with biological life.

Each of the above elements can be used separately or in combination to augment water's structure and to create *full spectrum living water*. This chapter begins the "dance." There are no strict rules—just the basic steps, the energy of the tools and the creativity of the dancer. Let your intuition and imagination guide you as you select methods that resonate with you and your circumstances. Above all, remember to connect with the water and to enjoy the dance!

Selecting a container

Plastic has become the universal container for everything. It is versatile, durable, lightweight, and inexpensive, but it is *not* the best container for water. Plastic contains toxins that may be gradually released in water. The type of plastic, its age, and the environment in which it is kept are all factors that may influence its effects on water. The extra energy in structured water can pull more toxins from plastic. This is especially true when water has been structured with paramagnetic substances, which tend to open the molecular pores in plastic. Plastic also creates "static-like" energy that interferes with water's natural electromagnetic field. It limits the developing *energy field* around a vessel of water. Whenever possible, avoid the use of plastic to store water—especially structured water.

Both glass and ceramic support water much better than plastic. Glass is made of silicon dioxide (quartz) the most common mineral on the Earth. Quartz and water have a similar molecular structure, and even though glass is not crystalline, it helps to support the molecular structure of water. It also carries the resonance of the Earth. Colored glass—particularly green, blue, and violet further sustain water's structure. Miron violet glass, developed in the Netherlands, blocks all visible light except the violet and infrared rays that are most supportive of water's structure. Tests show the preservation of life force when food and other preparations are stored in Miron violet glass.[2]

Ceramic materials are made of clay, which is also composed of silicate minerals. Clay has a crystalline structure—also similar to water's crystalline geometry. It holds the resonance of the Earth more strongly than glass and allows water to breathe, even when the clay has been glazed (although to a lesser degree). Ceramic materials protect water from long periods of direct light. Their porosity keeps water cool while they energetically refine water's structure. Clay is the best material for a water container.

Metal (particularly stainless steel) is often used for the storage of water. Even though stainless steel is better than plastic, metal is not a great container for water. Metals have an abundance of free electrons. More than four or five hours in a metal container and water loses considerable structure and life force. However, the Ayurvedic tradition of drinking from a copper cup has merit under certain conditions. Copper, silver, and gold are "oligodynamic." They have antimicrobial properties when placed in water. Copper and silver have the strongest oligodynamic effects. Drinking water from a copper cup (where the water has been allowed to sit for up to four hours) can be energizing and antibacterial. It is not recommended, however, where the concentration of copper in the source water is high, and it is not a good material for long-term water storage. Most silver vessels are alloys—also not recommended for water storage.

Preconditioning with prills

Prills are made of magnesium oxide (MgO), which has been used for many years to treat wastewater. MgO is a safe way to alkalize during water treatment (to prevent corrosion and eliminate odors in water caused by hydrogen sulfide). More occurs when MgO is added to water. Although MgO is only slightly soluble in water, the highly charged oxide ions are strongly attracted to water molecules, which causes gentle ionization—the first stage in the structuring of water. Prills gently break up the water network, helping to release old imprints. They are an excellent water pretreatment. Prills also slightly raise the pH in a reaction that gradually changes MgO to magnesium hydroxide (MgOH). Because both are white solids, there is no visible change to the prills during the process.

$$MgO + H_2O \gg Mg(OH)_2$$

Prills can be kiln dried to improve their resiliency when used in water. Typically sold in 3-oz. bags, kiln-dried prills can be used as a stand-alone water treatment or as a pretreatment prior to more thorough structuring. A 3-oz. bag of prills will treat a gallon of water in 12 to 24 hours. Prills may also be left in a larger container for a longer period of time. The authors use a bag of prills to precondition water for two to three days in 5-gallon glass containers. When used continuously, their effectiveness is gradually reduced; they should be replaced every 9 to 12 months. Prills can also play a role in the remineralization of distilled and reverse osmosis water (see page 168).

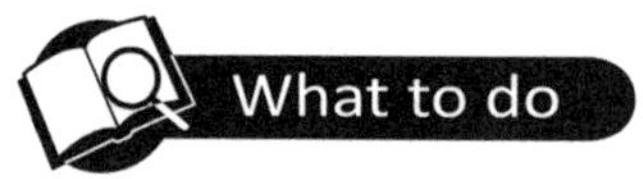

To use prills:

Add a 3-oz. bag of kiln-dried magnesium oxide prills to a gallon of water. Leave overnight or longer to precondition water prior to structuring. This step is not absolutely necessary, but it is an easy way to clear old imprints and jump-start the structuring process.

Adding minerals (salts)

Mineral balance is as important for water as it is for the organisms that consume it. Minerals anchor the life force in water, and they enhance the pathway for the flow of energy. It is well known that water without minerals does not conduct electricity. Minerals in ionic form provide free electrons for water to conduct electrical energy. Mineral ions also attract water molecules. They participate in water's structural stability[3] and in the creation of intricate geometric patterns capable of holding an infinite amount of information. Without ionic minerals, water cannot attain its highest potential. They are easily added as unprocessed "salts."

As a combination of two simple ions, salts contain a positive and a negative charge. They easily dissociate (come apart) in water. The addition of salts releases minerals in ionic form, enhancing any water prior to structuring, regardless of whether the source water already has minerals. There are many choices for adding ionic minerals to water. A variety of options are outlined below.

Unprocessed salts

Natural, unprocessed salts come from the ocean or an inland sea. They contain a balanced blend of essential minerals not available in all water. Salts include chlorides (a variety of minerals attached to the chloride ion), phosphates (minerals attached to the phosphate ion) sulfates (minerals attached to the sulfate ion) bicarbonates (minerals attached to the bicarbonate ion), and others. All the unprocessed, natural salts are high in chlorides, yet they contain some of each of the above. On the other hand, table salt contains only sodium and chloride—just two ions. It is imbalanced and unsuitable for adding to water—just as it is unsuitable for consumption.

There are many natural, unprocessed salts to choose from, including Himalayan salt, Celtic sea salt, Hawaiian salt, French Grey salt, Real Salt, and others. Each one has a different array of trace minerals and a unique energetic signature. Any of them will balance the minerals in water prior to structuring.

Note: Although the oceans have been severely compromised, the warm, moist environment in a salt holding pond causes salt to ionize and release an abundance of negative ions. During the time it takes salt to crystallize prior to harvest, most contaminants have been neutralized or oxidized in the sun. Beyond that, the natural ecology of a healthy salt pond keeps contaminants to a minimum. For the most part, naturally harvested sea salts are still safe for consumption.

Bamboo salt

Bamboo salt is an ormus-rich salt made in Korea. The process was developed by Buddhist monks over 1000 years ago to create mineral-dense salt with therapeutic qualities. The process begins with sea salt from the Korean Bay. It is packed into bamboo stalks and sealed with local clay. Salt-filled bamboo stalks are then roasted in clay ovens. The process can be repeated numerous times; each time, nutrients from the bamboo and clay are assimilated into the salt. Repeated roasting also concentrates ormus (see page 34). Originally, bamboo salt was roasted two or three times. Later, a Taoist healer named Insan (1909–1992) determined that nine times completed an alchemy, leaving the salt with a purple hue and a rich sulfur taste. The addition of bamboo salt to water helps to raise other minerals to their high-spin (ormus) state and to make minerals more biologically available.

Structured water made with bamboo salt that has been roasted nine times, is high in sulfur and iodine. Sulfur is the third most abundant mineral in the human body based on percentage of total weight. A large number of individuals are deficient—especially the elderly and the vegetarian community. Iodine is especially important to protect the thyroid gland from radioactive compounds. The minerals in bamboo salt are concentrated, so less is needed for structuring water.

Seawater

Microfiltered seawater is the premier salt solution for structuring water. The highly organized crystalline geometry inherent in this living matrix helps it to hold vibratory information better than any other solution. It is also capable of retaining structure longer. In addition to the complete list of minerals, seawater includes many important biological factors (see Chapter 4). Microfiltered seawater may be diluted to make an "isotonic" solution with the same saline concentration as human internal fluids. It is an ideal electrolyte replacement. To make an approximate isotonic solution from seawater, add two parts seawater to five parts springwater. Two well-known brands of microfiltered seawater are Ocean Plasma and Quinton's Marine Plasma.

Ocean mineral concentrates

OmniBlue Ocean Minerals and Polar Mins are two brands of ocean mineral concentrates that work well to remineralize and structure water. OmniBlue is harvested from the southern Antarctic Pacific currents of the Great Barrier Reef. Polar Mins is harvested from the Antarctic Ocean near the south magnetic pole. As concentrated liquids, both are low sodium/high magnesium—ideal for those who need to watch sodium intake. They provide a rich, mellow flavor to water.

Fulvic acids

Although water is not the largest source of minerals for the human body, Nature provided a way for the minerals to be biologically available. Fulvic acids hold minerals in organic complexes, taking them into the body in a form that is easily recognized. Fulvic acids are a natural part of the water in rivers and streams. They are also concentrated at the ocean floor. Unfortunately, fulvic acids are removed from water as "impurities" prior to water treatment (see page 59). When they are added, with unprocessed salts, the minerals in water become more biologically available.

Plant-derived mineral blends

Many mineral solutions can add a balance of ionic minerals to water. Some are extracted from ancient plant deposits and contain fulvic acids, and others may come from the sea. Look for the full complement of minerals in ionic form.

Add salts in ionic form for best results

It takes time for minerals to be assimilated into a developing crystalline array—especially in demineralized water (see Chapter 13). When minerals are added in liquid form, the process proceeds more rapidly. The authors of this book have noticed that the addition of dry salt results in a metallic taste to the water unless considerable time (days) elapses before consumption. On the other hand, the addition of the same salts, added in a saturated solution, produces water with a soft, rich taste.

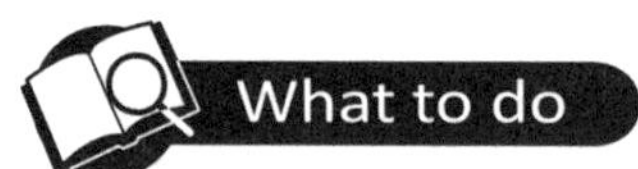

To add minerals in their ionic form:

Make a saturated salt solution according to the directions below *or* add any of the liquid solutions mentioned previously.

How to make a saturated salt solution

1. Begin with natural, unprocessed salt. Select salts that have not been heated or refined.
2. Add twice the amount of water to a volume of salt. (For example, add 1 cup of water to ½ cup of natural, unprocessed salt.)
3. Allow several hours or overnight for the salt to dissolve. (If all the salt dissolves, add more until salt remains undissolved in the bottom of the container. When no more salt will dissolve, the solution is saturated.)
4. You can either pour off the saturated salt solution and put it in a separate container or you can use the liquid from the top when adding it to water.

There will always be a certain about of dark-colored "rock" that will never dissolve. Discard it or let it remain on the bottom.

How much to add

The amount of salt solution to add depends on your water. If your water is from a natural source (spring or well) it will already have some minerals. In that case, add the smaller amount from the list below—or less. If your water source is distilled or purified, add the larger amount.

Saturated salt solutions made with unprocessed salt – Add ¼ to ½ tsp. per half gallon of water.

Bamboo salt – Add 2 to 10 drops of a saturated solution per half gallon.

Seawater (microfiltered) – Add 1 to 2 tbsp. per half gal.

OmniBlue or **Polar Mins** – Add 10 to 20 drops per half gal.

Fulvic acids – Add 5 to 30 drops/half gallon depending on the concentration of the product. The authors prefer to add fulvic acids 10 to 15 minutes prior to consumption or to add them to a container with a wide opening that can be easily cleaned. The organic acids leave a residue that builds up on the inside of a container, which requires more regular cleaning.

Other mineral blends – Add an amount that depends on the concentration of the product.

The amounts indicated above are designed to support water's structure. They do not represent therapeutic amounts for those interested in mineral supplementation.

Energizing your salt solutions

When ions are energized, they are more efficient carriers of energy. Thus, when a salt solution has been energized before it is added to water, the resulting water is enhanced to a greater degree. To energize your salt solutions, keep them over a triskelion (see page 159), inside a Tensor Ring (see page 156), or on/around orgonite (see page 155).

Structuring: a two-step process

The development of water's liquid crystalline structure is a two-step process that involves movement followed by stillness. During movement, the random, hydrogen-bonded water network is loosened, releasing dense vibratory information and old patterns; it prepares water for greater organization. The introduction of energetic input during this step (magnetic fields, sound, light, crystals, geometric wave guides, orgonite, etc.) supports the process. Some forms of energetic input are appropriate to use during both steps of the process.

Creating a vortex

Stirring or spinning is one of the easiest and most widely acknowledged ways to initiate movement for structuring water. Whether it is shaken, stirred, or sloshed, moving water always forms vortices, which induce tiny magnetic fields that contribute to water's developing structure. Vortices gather and concentrate energy. There are a number of ways to create a vortex—everything from manually stirring to the use of special devices.

Manually stirring

Stirring water is a wonderful way to create a vortex and to connect with the water. It's a great way to start your dance, even if you have nothing more than a pitcher and a wooden spoon.

To create a vortex by manually stirring:

1. Add ionic minerals to water in a suitable container and stir for several minutes in each direction. Speed (velocity) is not as important as connecting with the water while you stir.
2. After you have stirred in both directions for several minutes, you may want to further energize the water by stirring in a figure 8. The figure 8, otherwise known as the infinity symbol,

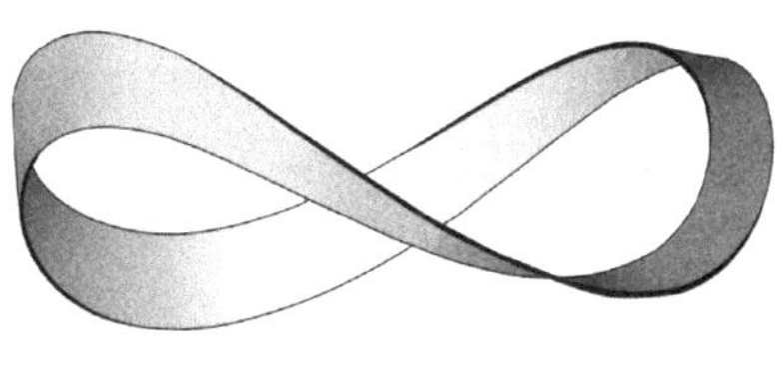

is used in many processes to energize and to repattern. The symbol is reminiscent of the mobius strip—a strip of paper that is twisted and attached at the ends forming a 2-D surface that can be traversed without end.

3. Another simple technique for manually stirring water is also an exercise for the brain. It provides *imploding* and *exploding* (feminine and masculine) energy for the water while balancing and integrating the right and left hemispheres of the brain. Some may notice an increased ability to focus when they do this exercise regularly:

   With your left hand, begin stirring the water in large circles in a counterclockwise direction. Gradually reduce the size of the circles until you are stirring only in the center of the container. Then slowly increase the size of the circles until you are stirring large circles again at the outer edges of the container.

   Reverse the direction of stirring so that you are making large circles in the clockwise direction. Gradually reduce the size of the circles, as before, until you are making very small circles in the center of the container—then increase the size of the circles again. Switch hands and repeat the entire process.

Selecting a stirring utensil

Wooden or copper stirring utensils, rather than stainless steel or plastic, are recommended. Copper conducts energy. This will help the structuring process. You can make your own stir wand from copper electrical wiring by intertwining two or three strands of wire. For an even more potent effect, use a compressed Tensor Ring (see page 156), referred to as a Tensor Wand (shown).

Adding magnetic energy

Water that is placed within a magnetic field becomes structured on its own (in time) even without stirring because magnets cause ions to spin. Stirring speeds the process. You can create a magnetic field by taping magnets around a pitcher as shown. Stirring water within this magnetic field is very energizing. The same method can be used on a blender—similar to the Volixer blender developed by Dr. Ron Cusson; however, the blades in blenders create very "cutting" energy that is not ideal for water.

Vortex Magnetizers

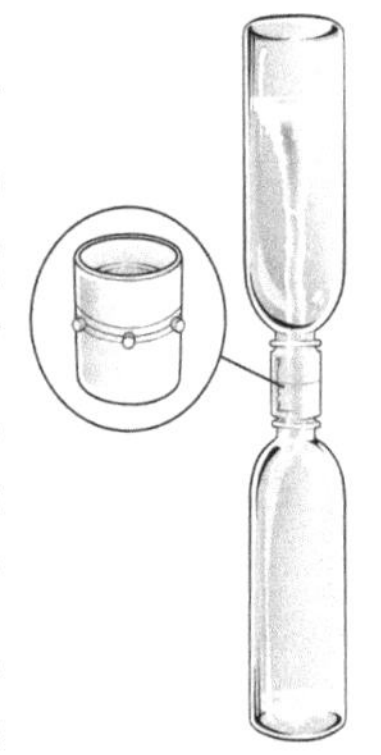

A simple device called a Vortex Magnetizer (VM) is an easy way to create a vortex accompanied by a magnetic field. The device connects two bottles as shown. Swirling creates a vortex, and spinning water passes through a magnetic field created by magnets in the VM. Each time water passes through the opening, greater coherence is established. The VM was designed to work with standard 2-liter plastic bottles, but many bottles will work—even some glass bottles. If you use plastic bottles, be sure to pour the resulting water into glass or ceramic.

Vortexing machines

Another kind of vortex generator automates the vortexing process. This type of device is about the size of a kitchen blender. Opposing magnets in the base provide a magnetic field while water spins. The device pictured is called the Tribest Duet Water Revitalizer. It releases a small amount of minerals from a basket in the base of the pitcher, but additional unprocessed salts are recommended—especially if you begin with demineralized (empty) water. (Several similar devices are marketed in many parts of the world.)

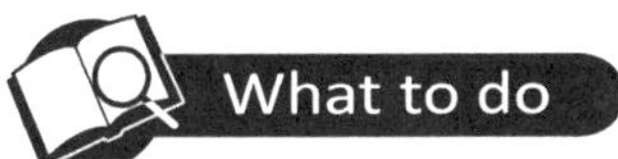

To vortex water using magnets:

1. Tape 8 magnets around a container as shown in the diagram on the opposite page, alternating north and south magnetic poles. Stir water within this magnetic field for 5 to 10 minutes, alternating directions every 2 to 3 minutes.
2. Fill a bottle ¾ full with water. Attach the Vortex Magnetizer (VM) to the empty bottle. Then screw the other end to the bottle filled with water. Hold the unit with one hand on the VM and the other hand on the top bottle filled with water. Swirl the upper bottle in a circular motion until a steady vortex appears in the center. Repeat the process at least 4 times—each time swirling in the opposite direction.
3. Use an automated vortexing device such as the Tribest Duet Water Revitalizer that incorporates a magnetic field to spin water for several minutes.

Adding magnetic energy with paramagnetic sand

Another way to add magnetic energy is with the use of paramagnetic sand. A small amount of paramagnetic sand is like having millions of tiny magnets interacting with the water. Placing paramagnetic sand *in*, *underneath*, or *around* a water vessel releases old patterns and brings structure to water. There are several ways to use paramagnetic sand to enhance your drinking water.

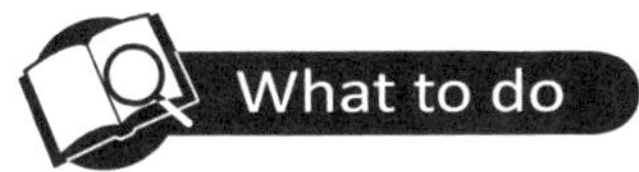

To structure water using paramagnetic sand:

1. Place paramagnetic sand in a bowl underneath your water. Leave for several hours or overnight.

2. Place paramagnetic sand inside a watertight enclosure and put it in your water. (The authors like to put paramagnetic sand inside a quartz tube sealed with silicone stoppers. Quartz participates in structuring the water, amplifies the resonances in the sand, and provides a watertight enclosure.) This method is ideal inside the holding tank of a crock or gravity water filter, where water remains for an extended period of time.

3. Surround water with paramagnetic sand. The easiest way to do this is to create a holding device with two tubes where the outer tube has a diameter about 2 inches larger than the inner tube. Fill the outer tube with paramagnetic sand. Place a bottle filled with water inside the inner tube. (The authors refer to this kind of device as an Earth resonance device.) Allowing the water to remain overnight brings profound refinement and Earth resonance to the water. This is similar to, yet beyond, the effects of MEOW devices, mentioned on page 216.

Flow forms create vortices

Flow forms are Nature's way of creating multiple interactive vortices. The term *flow form* refers to a variety of forms and shapes that allow water to find its own course while encouraging the inward curling that induces centripetal spirals and vortices. There are many types of flow forms—everything from large fountains that allow the water to drop into a series of bowls, to small handheld devices that are reminiscent

of waterfalls. Vortices are created every time water meets an obstacle, curling and spinning on its own, very much like water in Nature. When made of natural materials, these devices stabilize the water with harmonics from the Earth. Creating movement and vortices in this manner produces some of the liveliest (and lightest) water you can make. Augmenting the forms with magnetic and/or paramagnetic materials improves the outcome.

To structure water using a handheld flow form:

1. Use a waterfall-type device and pour water through several times.
2. Make your own waterfall-type flow form by cutting the end from a bottle or by finding a glass enclosure with a hole in the bottom. Fill the enclosure with a variety of stones and pour water through several times. Multiple passes allow water to release old patterns and to receive the energetic input from each stone in the collection. (See Chapter 14 for further information on specific stones to structure and program water.)

Fountains of Life

Some of the most distinctive vortexing devices we have seen are the creations of Randy Hatton. His handcrafted fountains generate golden ratio vortices as the water continuously circulates through magnetic fields and a variety of other enhancements. These exquisite works of art radiate energy far beyond the fountain itself. Simply watching the water is an uplifting and energizing experience.

Stillness brings maturity

Once water has been prepared via movement, it should remain still and in a cool, dark, quiet environment for several hours or longer. At this time, water molecules find their place in a crystalline array stabilized by minerals and the developing hydrogen-bonded network. Gas discharge visualization (GDV) analysis reveals that stillness after vortexing results in a measurable increase in life force.[4] As the hydrogen-bonded network stabilizes, water develops the creamy, smooth texture characteristic of *full-spectrum living water*. During this stage of water's development, water is more receptive than at any other time. It should be protected from outside electromagnetic fields and other forms of negative energetic input. Depending on your home environment, it may be best to avoid the placement of maturing water in a high traffic area.

Shape of the container

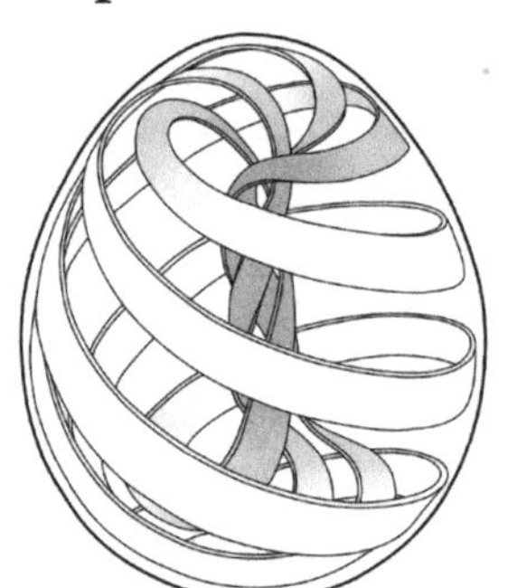

The shape and material of a holding vessel contribute to the development of water's liquid crystalline array. As previously discussed, geometry is the basis for the flow of energy in the universe. The geometry of a holding vessel can influence water's incubating life force.

Viktor Schauberger referred to the shape of the egg as a *life-motor,* in which "an enveloping, loving, and nurturing motion begins."[5] He explained that a life current is generated inside an egg-shaped container, detected as coolness at the narrow end and greater warmth at the rounded end of an egg.[6] Schauberger demonstrated how water in an egg-shaped vessel continues to cycle. It never becomes stale or stagnant. He created high-potency drinking water by placing the narrow end up.[7]

Numerous egg-shaped vessels have been developed for water. Many are made of ceramic materials. Unfortunately, most have been designed with the narrow end positioned at the bottom. This produces yang water—good for short periods of cleansing but not ideal for regular drinking. Yin water, full of the feminine essence, is produced when the, wide side of an egg-shaped vessel is placed closer to the Earth. Pictured is the Water Cradle—a unique egg-shaped vessel based on golden ratio proportion and made of laminar crystal (see page 151).

Another line of handcrafted vessels that incorporate golden ratio proportion is the Alladin family of carafes designed by an Austrian musician and engineer. Conceived in musical form—then converted to shapes that reflect golden ratio proportion—mouth-blown glass carafes produce a similar environment to that of an egg-shaped vessel. To further augment the design, each vessel has the Flower of Life pattern imprinted on the base of the carafe.

Although the geometry of an egg-shaped vessel or Alladin carafe is preferred, any container with a rounded bottom is better than a vessel with a flat or angular shape. Use what you have available; a quart jar will work while you look for a more desirable vessel.

Other ways to induce movement and stillness

Apart from using vortices and/or magnetic fields to cause movement, there are many ways to induce movement at a molecular level—even during relative stillness. When subtle energy is used instead of physical turbulence, the structuring process proceeds more slowly, but the resulting water is stable and refined. Many wavelengths of light and sound, orgonite (see page 155), and special geometric forms (including crystals) provide the energetic input required to loosen water's molecular network and to induce greater organization. These methods can be used *after* physical movement while the water is maturing, or they can be used *instead* of physical movement. The methods discussed next produce subtle vortices; they also add specific frequencies (information) to the water.

Crystals

In his 2005 literature review on the structure of water, Dr. Rustum Roy discussed the significance of a well-known but often overlooked phenomenon known as epitaxy.[8] Epitaxy is the transfer of molecular structural information from the surface of a material to a liquid—without the transfer of any of the material itself. In other words, by providing a structural pattern or template, it is possible to induce an entire body of water to *adopt* a predetermined pattern. Epitaxy explains why crystals bring structure to water. Their repeating molecular patterns based on the Platonic solids also make them natural transducers, capable of bringing universal source energy into 3-D form.

Another way crystals contribute to the structure of water is through piezoelectricity. The slightest pressure (like being surrounded by water) causes a crystal to generate weak piezoelectric current—another reason quartz crystals have been used for centuries to condition water. Any mineral in its crystalline form will structure water and leave its unique vibratory imprint in the water. A variety of crystals and other stones are discussed in Chapter 14 for their specific effects on the energetic qualities of water.

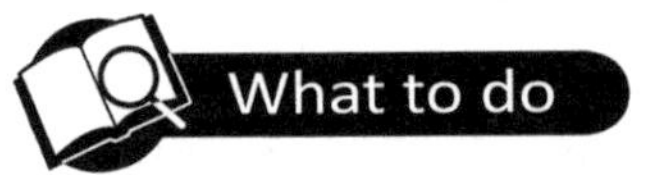

To structure water using crystals:

1. Place a crystal in a container of water and let it sit for several hours or overnight. Keep a quartz or other crystal in your water bottle to help maintain the structure of your water.
2. A specific type of quartz crystal, known as *enhydro* quartz,

works even more rapidly to structure water. These crystals contain a remnant of ancient water that has remained unchanged for eons inside the crystal—evidenced by tiny air bubbles formed as the water contracted inside the crystal. The structural organization of the water inside enhydro crystals is maintained by the crystal and translated easily to water. Place an enhydro crystal in water as in #1.

Laminar crystal

In the 1960s, what appeared to be a man-made tunnel was uncovered during excavation for a railroad. As with many tunnels discovered during excavation, it was sealed and abandoned. However, in this case, a man and his friend returned and opened the tunnel. Exploration revealed an array of artifacts, including a man-made material that looked like stone—made of mica and quartz embedded in ceramic. The material had a similar composition to artifacts found in numerous ancient ruins. When the *stone* was placed in water, some very interesting changes occurred to the water.[9]

Dr. Norman Shealy documented the discovery of the stone, which came to be called "laminar crystal." According to Shealy, placing laminar crystal in seawater produced oil-like water.[11] In other words, it structured the water.

The word *laminar* refers to *layered organization*. Organized layers amplify and concentrate energy. Water is a laminar substance. Its layered organization gives it the ability to store, and amplify energetic information. The Earth's minerals also form laminar assemblages as they crystallize—and crystals are well known to conduct, store, and amplify information. One of the most common laminar minerals on the Earth is mica. It has a number of unique properties that have made it very useful in the world of electronics. When it is placed around water, it amplifies the electromagnetic field around the water itself.

Laminar crystal is made today by combining clay with mica (and sometimes also with gold). When fired, a hard stone-like material results, similar to the *stone* created in ancient artifacts. Firing not only creates a resilient material; it also appears to spin many of the elements into a high-spin state—ormus. When placed in water, laminar crystal objects excite individual elements in the water and raise their energetic potential, which brings structure, energy, and many supportive frequencies to the water. Laminar crystal also attracts hydrogen from the atmosphere. In its high-spin state, hydrogen can remain stabilized in the water for a long time.

One of the first applications of modern laminar crystal was to "charge" other materials. Special chambers were created to charge the human body. At least one of these chambers is still in use today. (Dr. Shealy provides a regenerative spa experience that includes soaking in a hot tub, surrounded with mica, followed by relaxation on a bed of laminar crystal.)

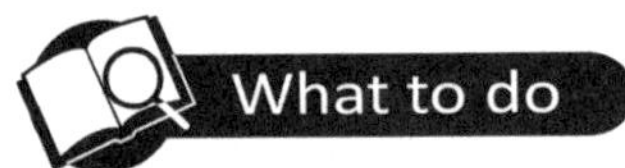

To structure water using laminar crystal:

Place laminar crystal objects (referred to as orbs, cosmic energy stones, pearls, or cupcakes, depending on who makes them) in a water container and allow the water to sit for 6 to12 hours depending on the size of the container and the size of the laminar crystal objects. When used in sets of three, laminar crystal objects set up a resonance that creates an even larger energetic field.

ANCHI Crystals

ANCHI Crystals are a blend of over 50 minerals, including tourmaline, mica, lepidolite, and quartz. Collectively, they carry a unique geo-imprint that has been found in only one place on Earth. The discovery of ANCHI Crystals occurred when two women sought a source of minerals to duplicate the mineral water found at many hot springs. To their surprise, when ANCHI Crystals

were added to baths and spas, profound effects began to be reported. The women were asked, "Do these minerals structure water?" They had no idea. They had never heard of structured water.

To answer the question, a sample of the minerals was sent to London. The answer came back affirmative—and later again confirmed at Penn State University.[12] Further investigation revealed that ANCHI Crystals communicated life force to water. This was demonstrated by Konstantin Korotkov, whose technique known as gas discharge visualization, was used to quantify the effect.[13] Continued work by Dr. Korotkov and a team of Russian scientists validated the energetic enhancements produced by the stones[14] (see Appendix A).

Study of the ANCHI minerals revealed that their effects on water were the result of a strong electromagnetic field.[15] Drinking water made with ANCHI Crystals and/or holding a small bag of the stones established a stronger and more coherent human electromagnetic field. Testing conducted by a variety of health practitioners revealed the balancing influence of ANCHI Crystals on the human body. Individuals balanced with ANCHI Crystals can, in turn, balance the electromagnetic field of water, plants, and other individuals via close proximity or touch.[16] In the authors' experience, holding water while wearing ANCHI Crystals creates ormus in the water, which explains the balance they confer to the electromagnetic field surrounding living organisms.

Beyond bringing structure to water, ANCHI Crystals transfer some of the Earth's original life-building patterns to water. Their energetic field is so potent that water is capable of holding these patterns for a long time. Sold in velvet pouches, ANCHI Crystals are very small and should generally be left in the pouch and placed *around* rather than *in* water.

To structure water using ANCHI Crystals:

Place a pouch of ANCHI Crystals at the base of a maturing water vessel or hang the pouch from the neck of a bottle. Leave it a few minutes to several hours or overnight.

Natural sound

Another beautiful way to gently vibrate water molecules into an organized matrix is with the use of sound. The sounds of Nature (wind, rain, water, crickets, bees, dolphins, etc.) and those generated by natural materials (crystal, rock, wood, the human voice, etc.) are excellent for helping water to release old patterns and to develop an organized matrix. Gongs, wooden flutes, crystal bowls, chants, and many of the sounds of Nature, gently vibrate water molecules and prepare the way for greater organization.

Crystal pyramids are a favorite of the authors. These pyramids, made of quartz, bring the resonance of the Earth, vortex energy, and sound together to add structure and coherence to a developing water matrix.

To structure water using natural sound:

1. Place water in an environment where it is exposed to the sounds of Nature or play the recorded sounds of Nature as water matures.
2. Play crystal bowls, wooden flutes, or sing/chant to the water.
3. Spin a "singing pyramid" over a vessel of maturing water. Let it spin clockwise and counterclockwise.

Orgonite

Orgonite is a term used to describe objects that encase layers of inorganic/metallic materials in an organic polymer (usually resin) for the purpose of concentrating life force and dispelling negative energy. The term comes from the word "orgone," coined by Wilhelm Reich in the 1940s. Orgone was Reich's term for life force. He discovered that by layering organic and metallic materials, life force could be concentrated. Reich built chambers he called *orgone accumulators* that were large enough for a person to sit in. Several hours a day relieved many symptoms.[17]

Orgonite is a more modern way of concentrating life force, based on Reich's discoveries. Because life force is a coherent, organizing energy, orgonite brings structure to water. It also protects water from the electromagnetic fields and other forms of negative energy present everywhere. The Internet is full of "how to" websites for making your own orgonite. The most important factor is to get as many layers as possible embedded in the resin. Here are other factors to consider:

- Smaller pieces of metal (especially metal powders) create the strongest orgonite because they provide more layers.
- Cones and pyramids are powerful shapes, but any shape will work. Larger is better, but a small piece made of a mixture of metal/carbon powders is more potent than a larger piece made with bigger pieces of metal.
- The inclusion of paramagnetic sand provides an additional organizing force for water.
- Mica flakes mixed in the resin amplify the energy and attract frequencies that support water's structure.
- The use of grounding materials (obsidian, shungite, hematite, etc.) energetically strengthen the orgonite and give it additional capacity to dispel "heavy" energy.

- Stones (crushed or otherwise) will add their unique energetic benefits. The inclusion of quartz crystals (especially at the top and outer edges) focuses and amplifies the energy emitted.
- The use of copper triskelions and Tensor Rings, enhances/ amplifies the energy of an orgonite piece. These can be embedded in the resin or used around orgonite—or both.

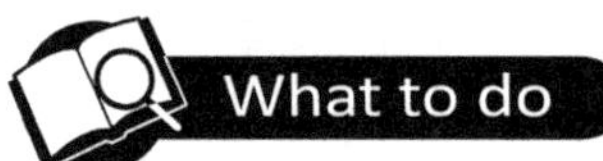

To structure water using orgonite:

Place a water vessel over an orgonite plate or near an orgonite piece to neutralize negative energy and accumulate life force as water matures.

Tensor Rings

Slim Spurling

A simple tool based on harmonics and geometry that brings structure and life force to water is referred to as a Light-Life Ring or Tensor Ring. This technology, invented by Slim Spurling in the 1980s and 1990s, creates a toroidal vortex, which can be visualized as a smoke ring. The energy is constantly folding inward to come around on the other side flowing outward. According to Spurling and others, Tensor Rings *potentize* water and can have a powerful effect on living biology.[18]

Spurling discovered that the ends of a length of wire have polarity (a positive and a negative side). If wire is looped back onto itself and joined together, one side of the loop has a positive energy, and the other side has a negative energy. To overcome this polarity, Spurling folded a length of wire in half. He also twisted the wire to negate interference. Then, each original piece of wire was connected back to itself, thus creating a Tensor Ring.

According to physicist John A. Wheeler, who outlined equations to explain the existence of tensor fields, "if you could see a tensor field, it would resemble the thin film that forms when a loop is dipped in a soapy liquid." Tensor Rings form what is called a tensor field across the opening of the ring. Those who perceive subtle energy describe a column of light that passes through the ring. The energetic column is an organizing force. It organizes the space within the column—including the molecular structure of water.

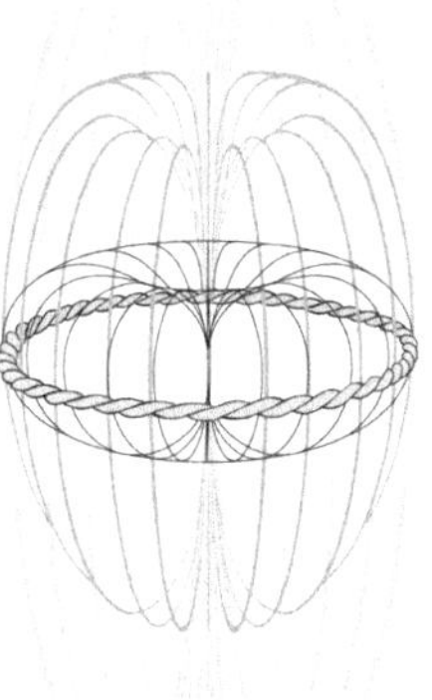

Hans Becker performed early research on Spurling's rings. He used a current probe with a special transformer to show that the rings amplified subtle energy over 400 times. He also determined (via spectrometry) that Tensor Rings allowed more light to pass through the area of the ring.[19] The energetic qualities of a closed-loop coil have sparked intense interest. According to Becker:

> When you create a closed loop coil, energy begins to flow and the laws of physics go out the window. They may be the simplest, most efficient source of energy there is—gathering it and condensing it freely from space. This is why planets follow elliptical paths and why everything from atoms to galaxies spins in circular motion. It is why the universe does not wind down and come to a stop.[20]

When making a Tensor Ring, the length of the wire is also important. Each wire is cut to a specific measurement. The first rings made by Spurling were based on the cubit length used to build the Great Pyramid in Egypt (20.6 in.). Later, other cubit lengths were identified. Brian Besco, who has continued to develop tools based on Spurling's work, has identified numerous cubit lengths—each with a resonance and properties of its own.[21]

Although Tensor Rings made of all the cubit lengths bring molecular structure to water, certain lengths are especially suited for water. In 2015, the authors identified a measurement of 19.65 in. with a resonance they refer to as the "Earth Core" resonance. Tensor Rings made using this measurement are tuned to the Earth's central core and to the Earth's natural energy field. They reconnect water (and anything placed in their energetic field) to the core energy of the Earth. This cubit length is extremely grounding and protective for water.

Tensor Rings are powerful tools. As water passes through (or sits within the energetic column of) a Tensor Ring, it becomes organized. The longer it sits inside the ring the more refined the structure becomes. The pH of the water is often balanced (this depends on the water source) and the taste of the water generally improves. Other improvements that are more difficult to quantify augment the water's energy. The authors use Tensor Rings during all stages of water preparation. They are especially useful for protecting water during maturation and storage. The column of energy produced shields water from outside interference. The authors have noted that Tensor Rings made of wire that is too large (with a gauge of 8 or less) are too powerful for water's delicate electromagnetic field. If you purchase or make a Tensor Ring, be sure that the wire is 10 gauge or larger. (**Note:** Wire gets larger as the gauge number gets smaller.)

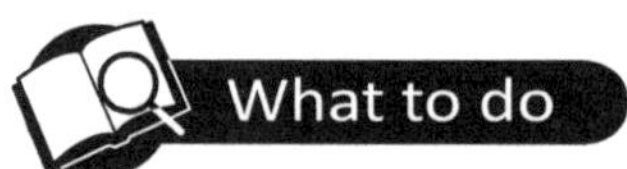

To structure water using a Tensor Ring:

1. Place a container of water inside the column of a Tensor Ring overnight or longer to structure the water and/or to protect it during maturation.

2. Pour water through a Tensor Ring several times.
3. Place a Tensor Ring around faucets, showerheads, and/or the incoming water line to your home.

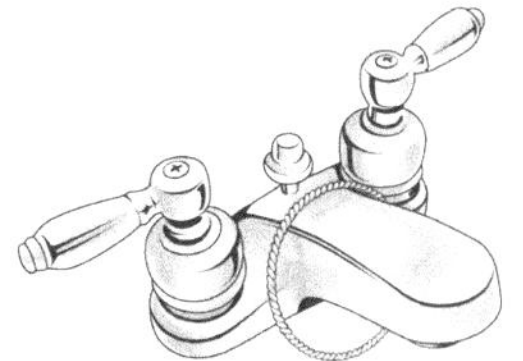

Triskelions—balance epitomized by the number three

The ancient symbol called the *triskelion* or *spiral of life* is based on the number three and the universal spiral. It is the epitome of balance. This symbol is found all over the world. It appears on Coptic antiquities, on the rocks of Mongolia, on Tibetan rings, on Buddhist banners, on ornaments of all the Himalayan countries, and on carvings of the Neolithic age. The symbol is drawn in one single line and represents the natural flow of the Earth in her seasons and cycles.

When constructed of certain metals, a triskelion sets up a bioelectric and biomagnetic energy flow based on mathematics and geometry. When the spirals touch each other, a circuit is completed that enhances the energetic flow. Placed around water, triskelions create vortices that bring structure to water. Used in conjunction with crystals and/or vortexing devices, they are a simple, yet powerful tool. The triskelion also has a unique ability to neutralize harmful energy in almost any setting.

Noticing the way twisted wire negated interference and changed the energetic output, the authors of *Dancing with Water* began to experiment with triskelions made of twisted wire. Three wires echoed the three spirals in the design and created a more delicate energy. Triskelions made with three twisted wires had a similar, yet gentler, effect on water. The authors refer to the triple twisted triskelion as the *feminine* version of the triskelion. They recommend both versions be used together for a more balanced energetic effect. They also found that when two of the spirals were placed along the same horizontal plane, the energy actually turned off. When they are slightly askew, the energy flows freely. (The triskelion at the beginning of this section is in the "off" position.)

Triskelions can be made of copper, silver, or gold—or a combination of these metals. The use of silver- and gold-plated copper wire is another way to work with masculine and feminine energy from a higher octave. A triskelion (made with two copper wires and one silver-plated copper wire) brings in the feminine energy of the Moon. A triskelion (made with two copper wires and one gold-plated copper wire) brings in the masculine energy of the Sun. Used as a pair, they work in a similar way (at a higher octave) to the copper pair to balance masculine and feminine energy for those who drink the resulting water. For instructions on making your own triskelions, see the video on the DancingwithWater.com website.

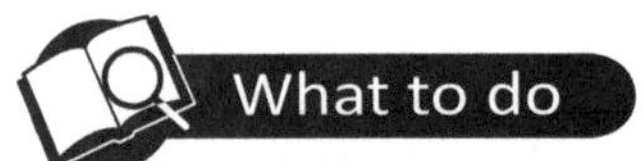

To structure water using triskelions:

1. Attach a pair of triskelions to a water vessel. Place one at the top of your water vessel in back. Attach its pair to the bottom on the front of the vessel. As water moves through the energetic field created by the pair of triskelions, it gradually releases old patterns and eventually assumes structure. (Notice how these triskelions are placed slightly askew for a better flow of energy.)

2. Attach a pair of triskelions to the bottom of a container (one on top of the other). The triskelion also helps to protect water from outside energetic influences.

3. Place a triskelion (or a pair) alongside or underneath your saturated salt solutions to keep them energized before adding them to your water. The triskelion is a nice complement to salt.

Flaska bottle

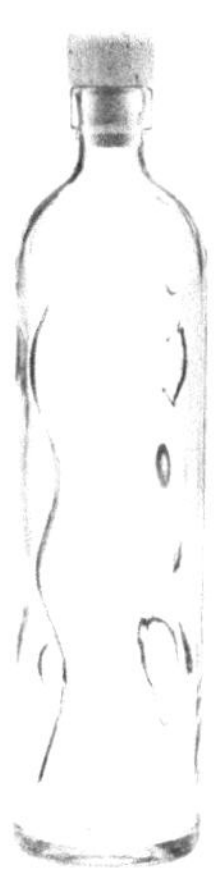

The Flaska water bottle, made in Slovenia, provides a unique way to surround water with vibratory information garnered from Nature and imprinted into the glass. Water placed in the bottle becomes structured in a short period of time. Flaska water bottles are imprinted using orgone technology. The glass bottles retain their imprint and ability to structure water for four years or longer depending on the extent of their exposure to electromagnetic fields. Evidence indicates that some water contaminants are rendered neutral when placed in the Flaska bottle. They are good for those who travel and for those whose only choice is tap water.

Plocher® Beverage Board

Plocher® technology from Germany imbeds biologically compatible frequencies from Nature into a wooden beverage board using methods similar to those identified by Wilhelm Reich. The effect is permanent. Water placed on the board for several minutes to overnight acquires the fresh, living quality of mountain springwater. Like the Flaska bottle, there is some evidence that even tap water is rendered biologically compatible. The board is lightweight and easy to transport; it can be used to energize and improve the flavor of many other beverages.

MRET technology

MRET (molecular resonance effect technology) devices were created by Igor Smirnov, who designed a crystalline polymer that generates subtle, piezoelectric vibrations when activated by an external magnetic field and oscillating light. Water produced via MRET carries frequencies that are very similar to the Earth's Schumann resonance.[22] These devices also provide the full spectrum of light frequencies from the Sun. The resulting water retains coherent structure for a long period of time.

A note about water ionizers

Electrolyzing devices, otherwise known as water ionizers, are reported to structure water. Although they do cause water molecules to align within the electric current, they do not bring lasting organization or coherence to water. Appendix B is a complete discussion of the pros and cons of water ionizers.

Structured water has unique and different physical properties,[23,24] giving it a greater capacity to conduct and store energy. This can make a difference in the health of the organisms that consume it. Although molecular organization is difficult to document, most people notice changes in how the water tastes and feels. Structured water is often described as "light," "smooth," "silky," and "creamy" however, not everyone notices the difference. Although you may not notice differences in the water itself, you may begin to notice differences in how you feel.

Image Credits (Chapter 12)

Page 149 Alladin Carafe: courtesy of Nature's Design. www.natures-design.com
Page 156 Slim Spurling: from the private collection of Cal Garrison
Page 161 Flaska water bottle: courtesy of Flaska; www.flaska.us
Page 161 Plocher Beverge Board: courtesy of www.plocher.com
Page 161 MRET Water Activator: courtesy of Igor Smirnov

References (Chapter 12)

1 Ho, M-W. (1998). *The Rainbow and the Worm: The Physics of Organisms.* 2nd ed. World Scientific, pp. 75-77.

2 Miron Violet Glass Website: http://www.miron-glas.com/en/Test-procedures

3 Oschman, J. (2003). *Energy Medicine in Therapeutics and Human Performance,* Butterworth-Heineman, p. 303.

4 Personal communication: Krishna Madappa following pre-posttesting of vortexed water. (2013). Data on file with MJ Pangman.

5 Schauberger, V. (1998). *Nature as Teacher* (translated and edited by Callum Coats), Gateway, p. 177.

6 Schauberger, V. (1998). *Nature as Teacher* (translated and edited by Callum Coats), Gateway, p. 177.

7 Bartholomew, A. (2010). *The Spiritual Life of Water*, Park Street Press, p. 178.

8 Roy, R. Tiller, W. Bell, I. and Hoover, M. (2005). The Structure of Liquid Water; Novel Insights from Materials Research; Potential Relevance to Homeopathy. *Materials Research Innovations*, pp. 577-608. Available online: http://ccrhindia.org/ijrh/3(2)/1.pdf

9 Shealy, N. (2000). *Holy Water, Sacred Oil: The Fountain of Youth.* Biogenics Books, pp. 1-17.

10 Shealy, N. (2005). *Life Beyond 100: Secrets of the Fountain of Youth.* Jeremy Tarcher/Penguin, p. 172.

11 Shealy, N. (2005). *Life Beyond 100: Secrets of the Fountain of Youth.* Jeremy Tarcher/Penguin, p. 188.

12 Swanson, C. (2008). Effect of Pomé/Anchi Crystals on Raman Spectrum of Water. Synchronized Energy Institute.

13 ANCHI Crystal booklet *Ancient ANCHI Crystals.*

14 Korotkov, K. (2002). GDV Research on Pomé-Anchi Minerals. Kirlionics Technologies. Research conducted for ANCHI Inc.

15 Swanson, C. (2008). Effect of Pomé/Anchi Crystals on Raman Spectrum of Water. Synchronized Energy Institute. Research conducted for ANCHI Inc.

16 Personal communication: Glynda Yoder, Co-founder ANCHI Crystals.

17 DeMeo, J. (1989).*The Orgone Accumulator Handbook: Construction Plans, Experimental Use and Protection Against Toxic Energy.* Natural Energy Works.

18 Garrison, C. (2004). *Slim Spurling's Universe.* IX-EL Publishing, pp. 41-42 and 51.

19 Personal communication: Hans Becker. (2009).

20 Personal communication: Hans Becker. (2009).

21 Besco, B. Twisted Sage Studios Website. http://twistedsage.com/

22 Smirnov, I. (2008). The Concept and Effects of MRET Activated Water, *Explore Magazine*, 17(6) pp. 62-64.

23 Pollack, G. Figueroa, X. and Zhao, Q. (2009). Molecules, Water, and Radiant Energy: New Clues for the Origin of Life. *International Journal of Molecular Sciences*, 10(4) pp. 1419-1429.

24 Chaplin, M. Water Structure and Science Website: www.lsbu.ac.uk/water/

Chapter 13

Remineralizing and Reconditioning Purified (Empty) Water

The pros and cons of consuming purified (empty) water have been debated for many years. Although many individuals are convinced that drinking "pure" water is healthy, Nature says otherwise. Water does not naturally exist without minerals (salts/electrolytes). Even rain that has been purified by the process of evaporation immediately begins gathering minerals on its journey back to the Earth. The internal fluids in our bodies are saline. Without salts, metabolism comes to a standstill.

The most common methods of purification (removal of everything from water) are reverse osmosis (RO), distillation, and deionization. The process of RO forces water through tiny openings that allow only water molecules to pass. Distillation pulls water into its vapor state, leaving contaminants behind. Deionization removes positively and negatively charged particles from water via ion exchange. Although these methods provide pure water, they leave it physically empty, aggressively hungry, and energetically sterile.

Although generally ignored, the consumption of demineralized water can have serious health effects.[1] Research conducted in Russia concluded that water low in minerals was a risk factor for hypertension and coronary heart disease, gastric and duodenal ulcers, chronic gastritis, goiter, and complications in newborns and infants, including jaundice, anemia, and growth disorders.[2] The German Society for Nutrition reached similar conclusions and warned the public against drinking distilled water. They reported symptoms that included cardiovascular disorders, fatigue, headache, and muscle cramps.[3]

Water is not considered a major source of minerals for the human body—but perhaps it should be. In today's world where food provides fewer and fewer minerals (and other nutrients) water is becoming a more significant source. Its role in maintaining electrolytic balance is well known. Why else would electrolyte drinks be recommended for mineral replacement after athletic exertion and illness? In the case of mineral deficiency, mineralized drinking water can play a relevant role.

True hydration does not occur with demineralized water. Perhaps this is why its consumption does not quench thirst.[4] The absence of minerals in water leaches electrolytes from the body causing water retention (water in the extracellular fluid outside the cells) and osmotic changes in the blood.[5,6] For these reasons and many others, minerals should always be added back to water after any form of purification. However, the simple addition of minerals *is not enough.* Purified water requires both *remineralization* AND *reconditioning* to bring it to a point where it can support organic life. Movement is critical, magnetic fields speed the process, and Earth resonance and beneficial microbes bring balance. This chapter will help you understand how to use salts, movement, magnetic fields, Earth resonance, and microbes to bring demineralized water back to a biologically compatible form.

Adding minerals

The addition of minerals in their ionized state is important—especially for purified water. In their liquid/ionized form, minerals find their place among water molecules more rapidly. Microfiltered seawater is the best salt solution for remineralizing purified water. It has the greatest capacity to fill the extreme emptiness of purified water. However, any natural, unprocessed salt solution will work.

You may also want to add extra bicarbonate salts, which are especially helpful to balance purified water. They provide buffering capacity (the ability to resist changes in pH) as well as a way for minerals to be delivered in their organic form. Bicarbonates provide numerous health benefits (see page 60).

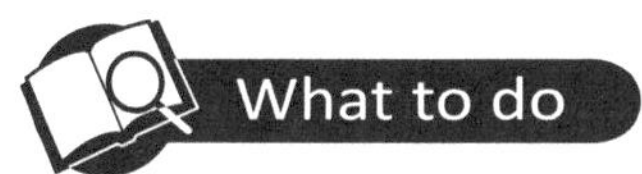

To add minerals to purified water:

1. Make a **saturated salt solution** or use any of the liquid salts outlined in Chapter 12. Add the *maximum* amount to remineralize empty water.
2. Make a concentrated solution of **magnesium bicarbonate** according to the directions on page 213. Add 1 tbsp. of the concentrate to each half gallon of purified water (optional).

Movement within magnetic fields

It can take days for purified water to assimilate minerals into the organized water matrix. Research conducted at Stanford University determined that purified water did not reach equilibrium for a number of days. However, placement of the water within a magnetic field hastened equilibration.[7] Remineralization requires movement—preferably within a magnetic field—and then a period of stillness, during which water is exposed to the resonant frequencies of the Earth. Water vortexing machines, such as the Tribest Duet Water Revitalizer, provide an optimum environment for the reconditioning of empty water. The repetitive movement within a magnetic field retunes each molecule, helping it to respond like a part of an *orchestra* rather than as an individual instrument (coherence). Manual methods such as the Vortex Magnetizer, and handheld flow forms that incorporate magnetics can accomplish a similar outcome; however, they require more attention.

To provide movement within a magnetic field, choose one or more of the following:

1. Run the **Tribest Duet Water Revitalizer** (or similar) on the 27-minute cycle. **Note:** The mineral basket with the Duet releases some minerals as the water spins, but to remineralize empty water, additional salts should be added.
2. Pour water through the **Vortex Magnetizer** 10 to 12 times.
3. Pour through a **handheld flow form** 10 to 12 times.
4. Attach magnets to a water container and stir for 20 to 30 minutes.
5. Place a crystal tube filled with **paramagnetic sand** in your water container. Allow the water to sit overnight or longer. This method takes longer, but the magnetics provide movement at a molecular level *and* a magnetic field.

Augmenting the mineral content during stillness

Other ways to augment the minerals in purified water are with prills, fulvic acids, and mineral stones. These can be present during, or they can be added afterward when the water is still. Consider any or all of the following:

Prills are good for reconditioning empty water because they eventually form magnesium bicarbonate in a two-step process.

$$MgO + H_2O \gg Mg(OH)_2$$

Prills slowly change in water to form magnesium hydroxide.

$$Mg(OH)_2 + 2CO_2 \gg Mg(HCO_3)_2$$

When magnesium hydroxide joins with carbon dioxide in empty water, magnesium bicarbonate forms.

Prills add magnesium and buffering capacity to empty water. Ideally, the amount of magnesium should exceed the amount of calcium in water, but this is rarely the case. Prills help to offset the imbalance. The longer they are left in empty water, the better.

Fulvic acids provide additional organic complexes to help carry minerals in a biologically available form.

Maifan stone has been used for centuries in Asia to add minerals and remove some contaminants from water. Maifan stone (also called medicine stone) contains nearly 70% silicon dioxide (quartz) and a balance of other minerals in their oxide form. Oxides react with water to form hydroxides (and eventually bicarbonates), which slightly raise the pH in otherwise acidic water. Allowing water to sit in a container with Maifan stone eventually adds a balancing variety of minerals in their bicarbonate form. Equally important, Maifan stone brings in Earth resonance.

To augment the minerals in water during stillness, use any or all of the following:

1. Add a 3-oz. bag of **prills** to a container of purified water. Let sit overnight or longer.
2. Add up to 1 tsp. of **fulvic acid** per half gallon of purified water depending on the concentration of the product you are using.

3. Add **Maifan stones** (up to ½ lb. per gallon of water). Let the water sit overnight or longer.

Earth resonance

Life on Earth is coupled with the Earth's magnetic field and its accompanying resonance, known as the Schumann resonance. Research has revealed the negative effects of isolation from this field[8,9]—the reason space stations include devices to provide the Earth's resonance. Isolation from the Earth's magnetic field has negative effects for water too. According to work conducted by Cyril Smith, water loses its ability to retain information (memory) when it is isolated from the Earth's geomagnetic field.[10] The vital connection with the Earth, which is severed by removing minerals, must be reestablished to allow water to carry life force once again.

There are a number of ways to reestablish the Earth connection. Most obvious is to place water directly on the Earth, where it can pick up the resonance of the Earth (and many natural sounds that resonate with Earth harmonics). Alternatively, the use of earthenware (clay) and glass (silicon dioxide/quartz) vessels bring in the Earth's energy, as does orgonite, laminar crystal, ceramic microbe technologies, Earthing devices, and many of the Earth's minerals. Any of the following will help to reestablish water's connection with the Earth:

Earthenware (ceramic) and glass containers resonate with the Earth. Ceramic holds the resonance of the Earth more strongly than glass and allows water to breathe.

Orgonite gathers life force and continuously dispels negative energy. It naturally establishes a connection with the organic life force present in and around the Earth (see page 155).

Laminar crystal carries (and amplifies) the resonance of the Earth while attracting many cosmic frequencies that are equally important for water (see page 151).

Beneficial microbes play a role in the natural development of *full-spectrum living water* (see page 113). Their presence balances the water terrain (similar to balancing the microbial flora within the human body) and also brings Earth resonance to water that will remain still for an extended period of time (overnight or longer).

Earthing devices bring frequencies of the Earth indoors through the grounding system that accompanies electrical wiring.

Paramagnetic sand and other materials amplify Earth resonance (see page 101). The authors of *Dancing with Water* have developed a special blend of Earth minerals that supplies both magnetics and Earth resonance for structuring and conditioning water. The blend includes paramagnetic sand, magnetite sand, mica flakes, quartz sand, and garnet sand. Placing this mix around a vessel of water brings in the resonance of the Earth and refines water's structure with a gentle magnetic field. It can also be used to fill quartz tubes that are placed around or *inside* a water vessel. Quartz magnifies the Earth's resonances and amplifies the influence of minerals in the mix.

Maifan stone also carries a strong Earth resonance. The addition of these stones not only adds beneficial minerals in their bicarbonate form, but reconnects water with the Earth.

To add Earth resonance, select any or all of the following:

1. Place water in a container of **glass or earthenware**. Allow to sit overnight or longer.
2. Place a piece of **orgonite** underneath or around an incubating water vessel. Allow to sit overnight or longer.
3. Place **laminar crystal** objects in water or place water in the Water Cradle. Leave overnight or longer.

4. Place several ceramic **QELBY balls** in a container of water. Leave overnight or longer.
5. Place water on an **Earthing** device. Leave overnight or longer.
6. Fill a quartz tube with the **paramagnetic, Earth resonance blend** and place it in a container of water, or make an Earth resonance device (see page 217). Leave overnight or longer.
7. Add a handful (up to 1/2 lb. per gallon of water) of **Maifan stones** to a container of water. Leave overnight or longer.

Guidelines for minerals in water

In 1980, the World Health Organization published the results of a study commissioned to provide guidelines for desalinated water. That study, conducted in Russia, concluded the following:

> Not only does completely demineralised water have unsatisfactory organoleptic properties [taste] but it also has a definite adverse influence on the animal and human organism.

The team of researchers recommended that demineralized water contain the following *minimum* levels of minerals:[11]

- Total dissolved solids—100 mg/l (ppm)
- Bicarbonate ions—30 mg/l (ppm)
- Calcium—30 mg/l (ppm)

Following the guidelines in this chapter exceeds the minimums established by this study while fulfilling the unestablished requirement for Earth resonance and the ability to carry life force, which is absent in purified water. The next chapter discusses how to add information to water.

References (Chapter 13)

1 Kozisek, F. (2005). Health Risks from Drinking Demineralized Water. National Institute of Public Health. Czech Republic. Chapter 12 in *Nutrients in Drinking Water, Water, Sanitation and Health Protection and the Human Environment.* World Health Organization Geneva. Available online: http://www.who.int/water_sanitation_health/dwq/nutrientschap12.pdf

2 Mudryi, I. (1999). Effects of the Mineral Composition of Drinking Water on the Population's Health (review, in Russian) *Gig Sanit*, pp. 15-18.

3 Distilled Water for Drinking? (In German) (1993). *Medizinischen Monatsschrift für Pharmazeuten*, 16(5), p. 146.

4 Kozisek, F. (2005). Health Risks from Drinking Demineralized Water. National Institute of Public Health. Czech Republic. Chapter 12 in *Nutrients in Drinking Water, Water, Sanitation and Health Protection and the Human Environment.* World Health Organization Geneva.

5 Robbins, D. and Sly, M. (1981). Serum Zinc and Demineralized Water. *American Journal of Clinical Nutrition*, 1(34), pp. 962-963.

6 Kozisek, F. (2005). Health Risks from Drinking Demineralized Water. National Institute of Public Health. Czech Republic. Chapter 12 in *Nutrients in Drinking Water, Water, Sanitation and Health Protection and the Human Environment.* World Health Organization Geneva.

7 Yamashita, M., Duffield, C., and Tiller, W. (2003). Direct Current Magnetic Field and Electromagnetic Field Effects on the pH and Oxidation-Reduction Potential Equilibration Rates of Water. *Langmuir*, 19(17), pp. 6851-6856. Available online: http://pubs.acs.org/doi/abs/10.1021/la034506h

8 Hunt, V. (1996). *Infinite Mind: Science of the Human Vibrations of Consciousness.* 2nd edition. Malibu Publishing, pp. 30-38.

9 Miller, I. (2013). Schumann Resonance, Psychophysical Regulation and PSI (Part I) *Journal of Consciousness and Research*, 4(6).

10 Smith, C. (2004). Quanta and Coherence Effects in Water and Living Systems. *Journal of Alternative and Complementary Medicine*, 10(1) pp. 69-78. Available online: http://citeseerx.ist.psu.edu/viewdoc/download?doi=10.1.1.605.3761&rep=rep1&type=pdf

11 Kozisek, F. (2005). Health Risks from Drinking Demineralized Water. National Institute of Public Health. Czech Republic. Chapter 12 in *Nutrients in Drinking Water, Water, Sanitation and Health Protection and the Human Environment.* World Health Organization Geneva.

Chapter 14

How to Add "Information" to Water

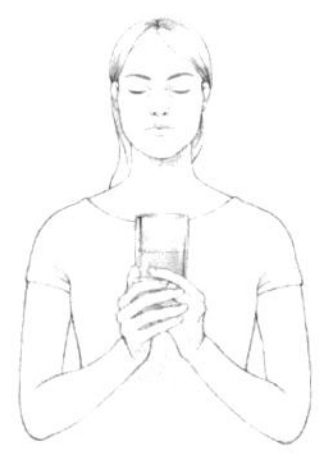

To structure water and then neglect to fill it with life force and vibratory information is a lot like drinking distilled water. It is energetically "empty." Water is a pliable matrix capable of adopting an infinite number of structural patterns. It can therefore be used to deliver therapeutic information.[1,2] This is the theory behind homeopathy, Bach Flower Remedies, flower essences, and gem elixirs. Adding information to water, sometimes referred to as *programming* or *imprinting*, is an important part of creating *full-spectrum living water*. The process involves exposing structured water to the frequencies or templates of the information you want to add. In some cases, this involves putting a substance *in* the water. However, physical contact is not necessary for water to pick up energetic information. Exposing water to the *energy field* of a substance under the right circumstances is enough for water to capture the information—especially if the water has been previously structured.

Structuring water by natural methods, such as those mentioned in Chapter 12, adds life force and sometimes adds information, but water is capable of holding and delivering much more. Why not take the dance to the limit? When you add specific information, you unleash water's full potential, and the dance takes on a new dimension. The ideas in this chapter are intended to get you started; the possibilities are endless.

Begin with structured water

Water that is structured is *ready to receive* information. Although many of the methods for imprinting water also provide structure for the water, if the water has been organized *first* it receives information more easily, and its ability to hold the information is strengthened.

Creating the right circumstances

According to Cyril W. Smith, who conducted significant research while at Salford University in England, imprinting water may be accomplished by proximity, contact, magnetic fields, shock waves (induced by homeopathic succussion), light, toroids, and vortical movement.[3,4] In the authors' experience, other *related* methods work too. Any of the following methods support the transfer of information to water.

Quartz crystals

Marcel Vogel (1917–1991), the IBM scientist who developed liquid crystal technology, discovered that many of the unique qualities of crystals resulted from their ability to organize and direct energy. Quartz crystals are resonators, transducers, and amplifiers of vibratory energy; they are used in many modern technologies because of these properties. According to Dan Davidson, the molecular lattice in a crystal acts as a container of information. It collects and holds energy that is eventually released in a *directionalized* manner, acting as a channel for the energy.[5] Because of their geometry and piezoelectric properties, quartz crystals are useful for directing and amplifying the information intended for water. They resonate at such a high frequency that they are better when counterbalanced with grounding stones.

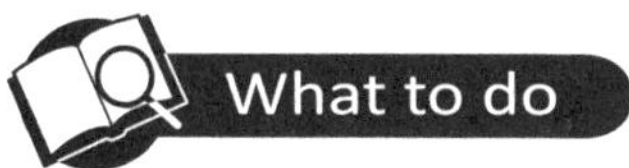

To imprint water using quartz crystals:

1. Place two quartz crystals on opposing sides of a glass or ceramic water vessel with their pointed ends directed toward the water. Counterbalance with two grounding stones as shown. (Grounding stones include black and green tourmaline, obsidian, smoky quartz, shungite, agates, hematite, and others.)
2. Position the material to be imprinted at the outer end of either (or both) crystals.
3. Lightly tap the container once, yet with enough force to create a slight reverberation within the water. This "sets" the imprint.

Magnets

Magnetism is fundamental in working with all living systems, including water. Smith proposed that the magnetic field "formats" water,[6] much like tapping the container in the procedure above. He identified a procedure for imprinting the frequency of an acupuncture point to water by bringing a pipette filled with water in contact with the acupuncture point, then bringing a permanent magnet close to the pipette.[7] Magnets are an easy way to imprint water. Even a refrigerator magnet will work if the volume of water is small. It should be noted that strong magnetic fields will also "erase" information from water, especially when the water is moving.

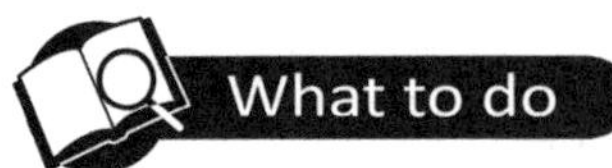

To imprint water using magnets:

1. Place the substance or vibration you wish to be imprinted underneath or to the side of a glass or ceramic container of water.
2. Bring a magnet up to the container and hold it for several seconds. It does not matter which pole of the magnet is brought to the water. Tapping the container sets the imprint even more strongly than simply bringing a magnet to the water.

Triskelion

A biomagnetic energy flow is generated by the spirals of a triskelion made of certain metals (see video on the dancingwithwater website). This tool works much like a magnet to amplify vibratory information and carry it into water. Triskelions can be used in the same way a magnet is used.

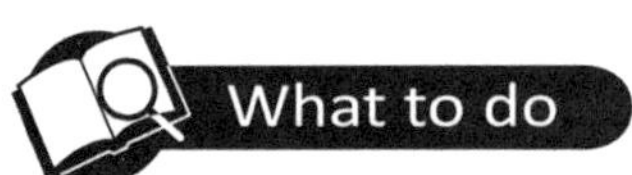

To imprint water using a triskelion:

Place a copper, silver, or gold triskelion, along with the substance or vibration you wish to imprint, underneath or to the side of a glass or ceramic container of water. Tap the container to set the imprint.

Tensor Ring

Tensor Rings create a toroidal vortex (a region of rotating movement where the flow takes on a doughnut shape).[8] Smith determined that toroids contain a magnetic field useful for imprinting water.[9] Tensor Rings also amplify subtle energetic input by 400 to 450 times, according to Hans Becker, who did much of the initial work with the rings.[10] Tensor Rings are invaluable when imprinting water. *All of the methods for adding information to water are enhanced with the use of a Tensor Ring.*

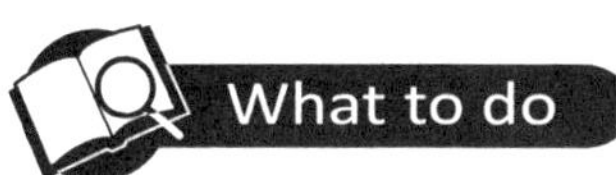

To imprint water using a Tensor Ring:

Place a glass or ceramic vessel of water in the energetic column of a Tensor Ring along with the material to be imprinted. Tap the container. You can also hold a Tensor Ring above or to the side of a vessel of water to imprint sound, light, or emotions from any direction.

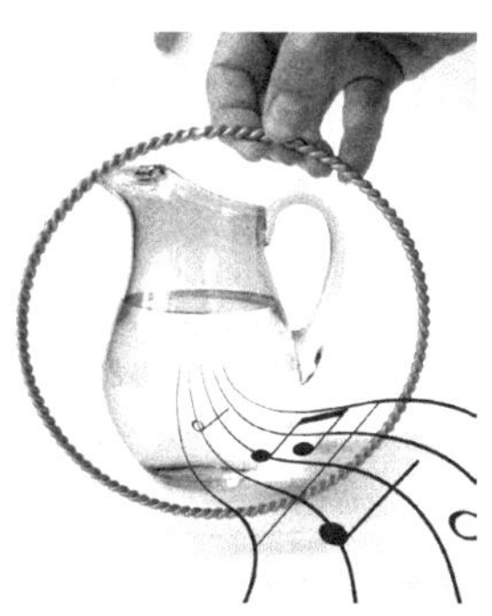

Light

Natural light is capable of carrying large amounts of information. As it passes through a substance, information is gathered, which can be directed into water. Lasers are the heart of some of the fastest methods of information transfer yet devised. Russian scientists have determined that lasers may be used to transfer information to blank CDs. When the imprinted CD is placed beneath a container of water, the information is transferred from the CD to the water.[11] Although they are powerful tools, lasers should be used with caution, especially when transferring information to water. If they are not used with clear intent, they can pick up the thoughts and emotions of the user, or energetic imprints from the environment.

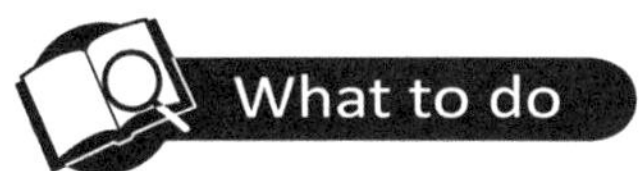

To imprint water using light:

1. To use natural sunlight, place the material for imprinting above the water container so that the Sun shines on it and

into the water. Tap the container lightly or leave for several minutes.

2. Using a laser pointer (it does not matter whether the light is white or has color), clear your thoughts, and focus your attention on the desired outcome. Direct the beam of light through the substance to be imprinted and into the water.
3. Clear your thoughts as in #2 above. Shine a laser through a substance onto a blank CD. The imprinted CD may then be placed beneath a container of water to transfer the information from the CD to the water.

ANCHI Crystals

ANCHI Crystals are accompanied by a strong electromagnetic field,[12] with some reports of Meissner-like properties.[13] Not only do these minerals imprint water with their own unique signature containing many of the Earth's original life-building patterns; their strong nonpolar magnetic field can imprint other vibratory information too. They are a library of Earth energy.

To imprint water using ANCHI Crystals:

Place a pouch of ANCHI Crystals next to the water container with the substance to be imprinted. Tap the container.

Paramagnetic sand

Paramagnetic sand also creates a nonpolar magnetic field resulting from the interaction of millions of paramagnetic particles. Anything placed with water while in the field will be imprinted. **Note:** Magnetite sand also creates a nonpolar field, but in the authors' experience, the use of magnetite can be overpowering for water. A *mixture* of paramagnetic and magnetite sand can be used where the ratio is 6:1, respectively. Other materials, such as mica and quartz sand, will enhance the mix.

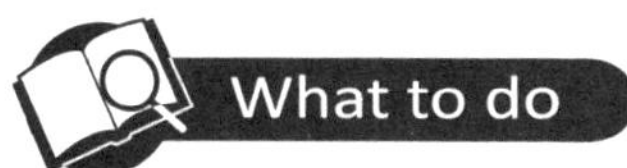

To imprint water using paramagnetic sand:

1. Place ½ cup or more of paramagnetic sand (or a blend as outlined above) in a dish made of a natural material (glass, ceramic, wood, etc.) along with the substance to be imprinted. Tap the container.
2. If you are using an Earth resonance device (see page 217) to mature your water, leave the material to be imprinted on top of the paramagnetic mix during the process. Because of the amount of time involved in this process, tapping is not necessary.

Orgonite

Orgonite can be a potent tool for transferring information to water. When made with paramagnetic sand, quartz, and mica, it amplifies the magnetic components that transfer information to water. Orgonite also protects water from negative energy in the environment.

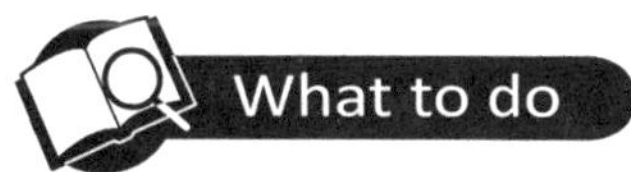

To imprint water using orgonite:

Place a vessel of water along with the material to be imprinted on an orgonite plate or adjacent to a piece of orgonite. Tap the vessel or leave for several hours.

Geometry

Certain shapes serve as wave guides, converting source energy into life force.[14] Pyramids and cones are very powerful collectors of life force and are good when placed over or around water. The shape of the egg is the most potent shape for water to be placed *within.* When water is placed inside an egg-shaped container (with the wide end down), it achieves its most receptive state, easily picking up the energetic milieu of its environment. **Note:** With the *pointed* end downward, even egg-shaped containers do not produce receptive water. In this orientation, the energy is always leaving and flowing back into the Earth. Water in an egg-shaped vessel with the pointed side down does not receive nor hold information well. It produces water good for periodic cleansing but not for *adding information.*

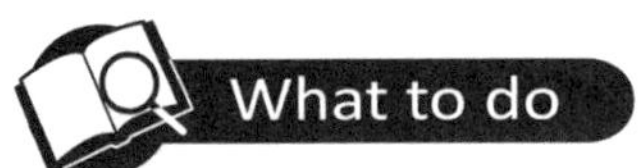

To imprint water using an egg-shaped vessel:

1. Place water in a properly oriented (wide side down) egg-shaped container (glass or ceramic) surrounded with a Tensor Ring.
2. Place the material to be imprinted inside the energetic column of the Tensor Ring and tap the container—or simply allow the water to mature. The Tensor Ring serves multiple purposes. It amplifies the signature of the material being imprinted, and at the same time, it structures and protects water from "outside" environmental influences while the water matures. Whenever you place water in a properly oriented egg-shaped vessel, you should protect it with a Tensor Ring and/or other protective support (orgonite, shungite, triskelion, obsidian, etc.). Pictured is the authors' creation, called the Water Cradle, which sits on an orgonite base with a Tensor Ring. Information from the stones on the base are being imprinted into the water.

Platonic solids

All crystalline structures are based on patterns in Platonic solids, distinguished by phi proportion. The Platonic solids provide universal patterns that contribute to coherence. When they are introduced during the imprinting process, they amplify the water's ability to hold any imprint.

To use Platonic solids during imprinting:

Place any or all of the Platonic solids, preferably made of quartz or other appropriate stone, into the water being imprinted. Their presence strengthens the imprint.

Granite

Granite is the most abundant basement rock on Earth. It contains quartz, mica, and other minerals that amplify signals in the Earth. Granite does not work as quickly as many other methods of imprinting water, but it is exceptional for bringing the Earth's resonance. Water placed on a slab of granite receives Earth resonance and the amplified frequencies in the immediate environment. **Note:** Some granite can be radioactive. Outdoors, natural radioactivity is dissipated, but if you decide to bring granite inside, make sure the granite does not contain radioactive elements.

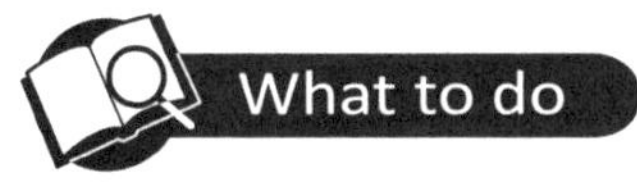

To imprint water using granite:

Place water and the substance you want to be imprinted on a slab of granite or on a granite countertop. Leave for at least 8 hours so the water receives the imprint, which is supported and strengthened by the Earth's resonance.

Human crystals

When you are focused and your energy is balanced, you become a powerful receiver and transmitter of life force energy. Your focused attention can be just as potent as any of the methods listed previously. Particularly when the body's water is structured, the human crystalline form is the most powerful crystal there is.

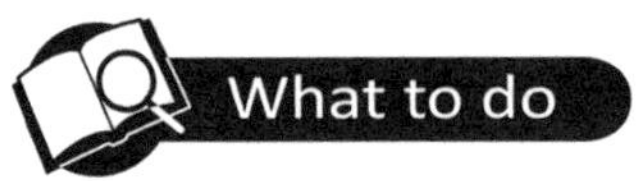

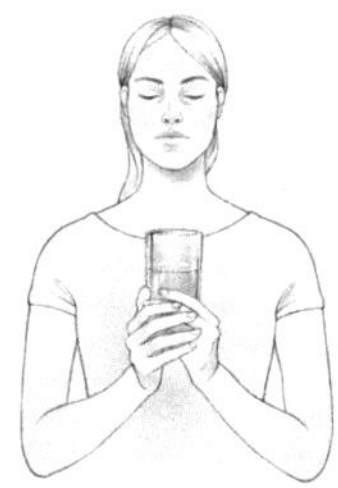

To imprint water using focused attention:

Focus on the energy you want to be imprinted in the water and see/feel/imagine the transfer of energy. The keys to this method are balance, focus, and your heart-centered belief in the process.

Kinds of information to add

Everything has a frequency/vibration, with the potential for bringing balance or imbalance. The natural world is filled with supportive vibrations; it is one reason spending time outdoors is so regenerative. Water picks up vibrations from its environment. If those vibrations are from the natural world, the water will keep us in harmony with the Earth and all life on the planet, but if water carries the vibrations of treatment chemicals, plastic, unnatural electromagnetics, and man-made materials, it cannot support healthy life. When selecting *information* for your water, natural is better. For example, there is a big difference between the electronically generated frequency of a note than the same tone found in Nature. Singing the note yourself may be even better. The remainder of the chapter includes ideas for adding information to water from minerals and plants, sound and music, light and color, words and symbols, and thoughts and emotions.

Adding information from the mineral kingdom

The Earth's rocks and minerals are full of vital energy. Imprinting your drinking water is similar to wearing stones in your energy field—better in many ways because the vibrations are delivered deep into your cellular matrix. Many books have been written about the healing and awakening properties of gems and minerals.[15,16,17] Almost any stone can have therapeutic properties, but some are particularly potent.

The next section introduces a few stones that have a special resonance with water. Some stones are better when placed directly in water, but others release toxins in the aqueous environment. Also, make sure the stones you use are energetically and physically clean. The easiest way to energetically cleanse a stone is to place it in alternating hot and cold running water for several minutes (back and forth several times). Many stones can be placed in the sun. This method of cleansing is preferred when heavy energy needs to be released. You can leave them for several hours or for several day/night cycles depending on the energy to be cleared. (Aquamarine, rose quartz, amethyst, citrine, and some others should not be left for more than several hours because the sun may bleach their color.) You can also energetically cleanse stones by placing them in a bowl on top of unprocessed salt overnight or in water with a rough diamond.

Water-related stones (alphabetically listed)

Aquamarine – Drinking aquamarine water is like sitting by a waterfall. It calms emotions, especially volatile emotions. Aquamarine also helps people express their feelings—particularly those who have been shut down by trauma or who are not comfortable *feeling* their emotions. It connects the heart and mind. Aquamarine water can also have an impact on physical conditions that are typified by rigidity and calcification (arthritis, arteriosclerosis, stiff joints and muscles, etc.).[18] Aquamarine may be placed directly in water.

Azumar – A new stone, discovered in 2013, azumar awakens ancient codes for joyful living. It carries wavelike pulsations that organize water's energy field and deliver codes for serenity, confidence, and joy. According to Robert Simmons, author of *Stones of the New Consciousness*, azumar "transforms the body's inner ocean to a moving sea of rhythmic bliss.... Its currents wash through the Liquid Crystal Body Matrix, bringing refreshment, joy and rejuvenation to every cell." Azumar is also intimately connected with the consciousness of water. Its discovery comes at a time when many are making a conscious connection with water, and it can assist in the process.

On a physical level, azumar supports changes in our evolving DNA and helps the body to assimilate transformational information being delivered by the Earth and the Sun. On an emotional level, azumar supports the release of emotions, such as depression. It raises the frequency of the body's internal water so that long-held emotional patterns keeping a person stuck can be resolved and released. On a spiritual level, azumar connects us with the joyful flow of life. Azumar promises to teach us how to be like a drop of water—individual, yet "at one" in the ocean of life. **Note:** Some pieces of azumar may contain an occasional bit of stibnite (gray color) that is toxic when placed in water. It is best not to place azumar directly in water.

Blue topaz – Blue topaz carries the blueprint for water's original life-sustaining patterns. Placing a tiny piece of nonirradiated, gem-quality blue topaz in water will restore the water's original patterns and its structure within seconds. Pieces that lack gem quality will structure water in several minutes. **Note:** Using irradiated blue topaz will not structure water. Unfortunately, most blue topaz on the market today is irradiated to enhance its color. Blue topaz may be placed directly in water.

Diamond – Diamonds have an extremely pure and powerful resonance tuned to higher consciousness. When placed in water, rough diamonds bring structure and the full spectrum of light to water. Even very tiny rough diamonds will work. Diamond water also has a precise energy capable of cutting through energetic blockages. The energy is so strong that it can cause instability (on the mental level) if diamond water is used inappropriately. If you are considering the use of diamond water, be ready to let go of many previous perceptions of life—anything that might stand in the way of higher truth. Also, be aware that the use of diamond water will dislodge energetic blockages but that it will not *clear* them. You should have access to tools and/or facilitators who can help you release the patterns and programs that block your spiritual progress. If you are not ready for deeper cleansing, you should wait. Diamond water is not meant to be used indiscriminately. It should always be used with purpose. One way to "temper" the precise and cutting energy of diamond water is with the use of gold. When used with a diamond, gold helps the physical body to integrate the powerful energy. The safest and easiest way to use diamond water is to proceed in three phases.

- **Phase 1** Place a rough diamond in a bowl of water in your living space; a water fountain is ideal. This water will keep your living space clear of negativity. The energy will prepare you for further exposure to diamond energy.
- **Phase 2** Bathe in diamond water. To make a diamond water bath, prepare a gallon of diamond water (place a diamond in the water for about 8 hours), then pour this water in your bath with other selected salts. Soak for about 30 minutes, following your inner guidance. Diamond water baths are also appropriate before giving birth and prior to other important occasions.
- **Phase 3** Drink diamond water—always with purpose. Place a rough diamond in a glass of water for approximately 10 minutes or longer. Have a very clear intention as you consume the water. Some people may never be inclined to actually drink diamond water, some may drink it only once, and others may drink it as they are prompted.

Diamond water is ideal for placement in the Earth's waterways, where it will help water to release the negativity of long-term disrespect and disregard. **Note:** Faceted diamonds that have been worn in jewelry are not suitable for making diamond water. They carry the imprints of the owner and/or the relationship they support. Diamonds may be placed directly in water.

Larimar – This watery blue stone (also called dolphin stone) is extremely calming and stress relieving. Larimar supports the emotional changes and Earth changes we are experiencing now as a human race. It is a beautifully matched complement to seawater. Larimar may be used directly in water.

Mica – Mica has energizing and cleansing properties. It is abundant everywhere on the planet and acts as a shield against discordant energy. Mica deflects chaotic disturbances that might interfere with the bioregenerative systems on this planet. All forms of mica are closely tuned to the resonance of the Earth and to the corresponding cosmic frequencies of life. Biotite (black) mica holds the program of Earth's primordial water—very helpful for returning water to its former state. Mica was used in ancient times in many parts of the world (and in conjunction with most of the pyramid structures) as a means of receiving and transmitting energy. It energizes the bioelectric field around cells, organisms, and large structures. **Note:** Mica should not be used directly in water. Its flaky layers may easily break off in the water.

Quartz – In its crystalline form, quartz has the same basic molecular structure as water. It plays a key role in the natural structuring of water on the Earth. Because of its high vibratory rate, it is best when used in conjunction with grounding stones (obsidian, hematite, mica, shungite, agate, smoky quartz, tourmaline, petrified wood, and others). Quartz is often used directly in water. In fact, quartz sand has been used for many years to "polish" water following filtration.

Rose quartz – This pink form of quartz is very emotionally balancing and stress relieving. Its purpose is to remind us that we are loved and to bring us into a state of *being* loved. Rose quartz water is wonderful for children. It may be used directly in water.

Shungite – Shungite is a unique combination of carbon and silicate minerals found in an ancient deposit in Russia. Its uniqueness stems from the presence of hollow molecular cages, known as *fullerenes,* embedded within the rock. When placed in water, shungite initiates the development of fullerene-like cages in the molecular structure of the water. It has been used in water treatment in Russia for many years. Shungite water also has many documented therapeutic properties, including antioxidant benefits, antihistamine properties, and energy-balancing effects.[19] Two kinds of shungite (sometimes referred to as black and silver) have differing amounts of carbon and fullerenes. Black shungite is common and relatively inexpensive. Silver shungite, also called elite or noble shungite, is more rare and has a much higher concentration of carbon and fullerenes. It structures water much more rapidly. Shungite is recommended to be used directly in water—and to remain there until the water is consumed so that the fullerene-like structure is maintained.

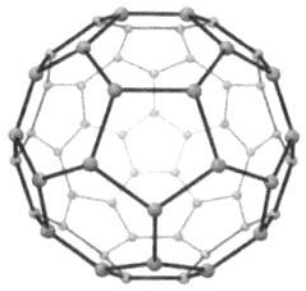

Tourmaline – Tourmaline carries far infrared energy—the energy of life. It also carries the energy of nourishment, especially the nourishment delivered in water. Tourmaline is a good stone to use in conjunction with all imprinting because it prepares cells to receive energetic nourishment. Research in Japan revealed tourmaline activates metabolic processes; it improves circulation, relieves stress, increases mental alertness, and strengthens the immune system.[20] Pink, green, and gold tourmaline may be placed directly in water. Black tourmaline is a good grounding stone and is best used *around* water as a protection and/or counterbalance for quartz.

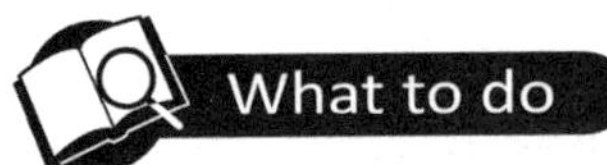

To imprint water using stones and crystals:

Use any of the methods from the previous section for imprinting the energy of stones in water.

Adding information from the plant kingdom

The system known as homeopathy is based on adding *information* to water. Remedies that include many of the Earth's plants are diluted to the extent that what remains is *only* the information and not the substance itself. Water is an ideal delivery system. Similar, yet different from homeopathy, the healing power of flower essences was discovered by Dr. Edward Bach. He created 38 flower essences with restorative properties known as the Bach Flower Remedies. Since the original 38 remedies, the healing traits of other flowers have been identified and described. It is best to use homeopathic remedies and plant/flower essences made by those with appropriate expertise. Their use in structured water made with the resonance of the Earth further potentizes them.

To enhance homeopathic remedies:

Add the recommended amount of a homeopathic remedy or plant essence to structured water, although you may need less to accomplish the same purpose. Tapping the container lightly on your hand further sets the imprint.

Adding sound and musical information

In Dr. Emoto's first *Message from Water* book, he released photos of water crystals that formed following water's exposure to different kinds of music. Music with a beat or vibration that was *out of sync* with organic life hindered its ability to form a crystal when frozen. Other music formed beautiful crystals.[21] Because the body is mostly water, imprinting water with natural sound and music is another way to provide nourishment throughout the day. You may want to reread Chapter 8 on sound.

Musical instruments

Each class of instruments has its own effect on biology. The same effect can be imprinted to water. Woodwind instruments tend to create a meditative and calming effect for turning inward. Brass instruments have a clearing and energizing effect. Stringed instruments set a more emotional tone, and percussion instruments (especially drums) carry the energy of intention. Instrumental music (preferred over electronically produced sound) can be very balancing for the human body. Imprinted in water, the effects are similar.

To imprint water with music:

Using your intention to help focus the vibration, imagine the music flowing from the source to your water. Support the process with a Tensor Ring or any of the methods above. Music does not have to be louder than normal, but it is better to have the source of music in the same room with the water. Original music is ideal, but recorded music works too.

The human voice

The sound of your own voice is extremely powerful. It can nourish your body in many ways. And if laughter is the best medicine, then laughing and giggling with the water you drink is medicine for your soul.

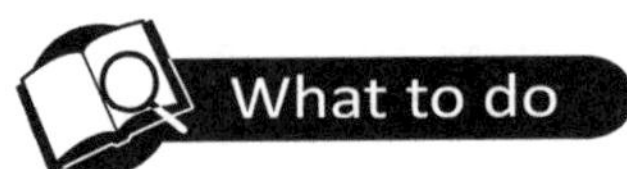

To imprint water using your own voice:

1. Speak affirmations to your water, including things that may support you throughout your day or through a troubling time. Use of the words "I am" can be very potent. They put a person in the energy of truth, for example, "I am loving and kind as I interact with those around me." Tap the glass as you speak, to set each imprint.
2. Sing and laugh with your water (gather children if you have them). You may be surprised by how energized and uplifted you feel when you drink this water.

Nature's vibrations

The sounds of Nature are wonderful to listen to, but there is something about receiving them through water that is even more nourishing. It is actually one of the ways Nature intended for us to receive nourishment. Like the frequencies of plants and minerals, the sound of waves at the ocean and the wind through the trees is both relaxing and balancing. Sounds generated by the animal/insect kingdoms are equally supportive (birds, bees, crickets, dolphins, etc.). And beyond the sounds of our own planet are the inaudible sounds of the solar system and the rest of the universe. The Earth has her own *hum*, referred to by the authors as Earth resonance. Our biology is especially tuned to this resonance. It has slight variations throughout the day, with seasonal variations we were meant to receive from water. Staying in tune with the Earth itself and the sounds of Nature helps you to stay in balance and to make seasonal transitions with greater ease.

To add vibratory imprints from the natural world to water:

1. Take a glass or ceramic vessel of structured water with you to the ocean, to the forest, to a flower bed while the bees are collecting pollen, or on a Nature walk. Place it directly on the

Earth for 5 minutes or longer where the desired sounds of Nature are present. Let the Earth set the imprint.

2. Play recorded Nature sounds to your water. There are many good CDs, but making your own recordings of the ocean, songbirds, bees, crickets, dolphins, thunderstorms, and other natural sounds is often better. They are projects you will never forget.
3. To imprint the 24-hour spectrum of the Earth's frequencies, place water in a closed glass or ceramic container directly on the Earth in a cool, dark place for 24 hours. (Caves are great places to imprint water with the frequencies of the Earth.)
4. Make an Earth resonance device (see page 217). Allow the water to mature for four or five days.

Adding information from light and color

Water was designed to carry the Sun's nutrients (electromagnetic spectrum) to all living things. Drinking water that has been placed in the sunlight is a wonderful way to get the vibratory energy of the complete color spectrum. It is also very energizing as long as the water has not been left too long in the Sun. Moonlight carries the light of the moon with accompanying information from the stars and planets in the night sky.

The Sun can also be used to imprint colors in water. Color therapy has been used for many years ever since the work of Edwin Babbitt, who developed the concept of healing with light and color. The most well-known way of imprinting color includes wrapping colored cellophane around a bottle of water and placing it in the Sun for several hours or putting water in colored glass jars. But there are more natural and more effective ways. When water is structured first, it does not take hours in the Sun to accomplish the task. In fact, hours in sunlight is counterproductive for reasons discussed in Chapter 7.

Indirect sunlight is better for infusing color than direct sunlight. Placing water in the indirect light beneath or around a plant or flower is a much gentler way. This infuses the water with a balanced blend of colors in a desired frequency *range*. It is also superior to having light penetrate through cellophane with synthetic colorings and substances in the plastic that may not resonate with the human body (all are carried with the wavelengths of light into the water).

Another way to infuse color (especially during the winter) is by covering a container of water with a natural fiber cloth dyed with plant juices. There are dozens of leaves, roots, seeds, and flowers that can make natural dyes. Even placing a glass of water *next to* a plant will have the desired effect because the colors are reflected into the nearby water. Cloth dyed with natural pigments that include DNA (which takes spirals of light into the water) goes far beyond cellophane to infuse color in water. You may want to reread Chapter 7 on light.

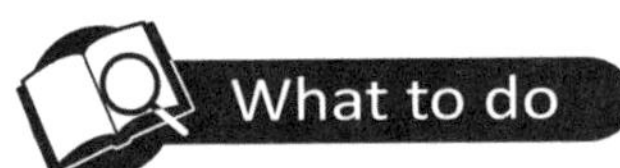

To imprint water with light or color:

1. To make Sun water with the full spectrum of the Sun's electromagnetic field, place a glass container directly on the Earth. In this way, the gentler Earth frequencies will balance the strong solar energy. Depending on the time of year, the Sun's position, and cloud cover, leave it exposed for no more than 10 to 30 minutes.

2. To imprint water with the energy of the moon, stars, and planets, choose a clear night (and remember that the phases of the moon and the position of the stars all have an effect on water). Place your covered glass container directly on the Earth in a place where it will receive the full night sky. The longer you can leave it out the better, but one or two hours will work. Remember to bring it in before the Sun comes up.

3. To infuse color in water, place water directly on the Earth beneath a plant or flower with the desired colors. Allow the water to sit undisturbed for at least 5 minutes out of direct exposure to the Sun. Alternatively, water can be placed next to a plant or flower on a windowsill. Allow the water to remain for several hours.

4. Dye a piece of natural fiber cloth with the juices of plant leaves, roots, seeds, or flowers to create a desired color. Cover your water with this cloth and allow it to remain for several hours in indirect light.

Words and symbols

Using sensitive equipment to measure magnetic fields, Dan Davidson determined that even line drawings cause changes in the flow of the source field[22] (see Chapter 6). A simple triangle drawn on paper showed concentrations of energy at the vertices. When the triangle was enclosed by a circle, energy was focused in the center of the triangle. His research validated the use of symbols to change the energy in an environment.

Dr. Emoto demonstrated that written words leave their vibratory imprint in water too. When placed on water containers, unkind phrases, such as "You fool" and "You make me sick," prevented the water from forming crystals as it was frozen.[23] In his experience, words of love and gratitude often overcame negative imprints in the water, allowing intricate crystals to develop during freezing.

In the 1970s, Malcolm Rae and a group of doctors developed Magneto Geometry—a way to convert the vibratory signature of a substance into a geometric pattern. Building on this work, Don Gerrard identified the geometric patterns from a variety of homeopathic remedies and published what he referred to as the *Vibrational Medicine Cabinet*. Although the book is out of print, it can often be purchased online.[24]

Beginning in the 1960s, Egyptian architect and natural scientist Dr. Ibrahim Karim, developed an advanced science called BioGeometry, based on the discovery of vibratory symbols incorporated in ancient Egyptian temples. The science uses the energy of shape (both 2-D and 3-D) to balance energy fields. It includes over 700 patterns, called "biosignatures" (natural patterns of energy movement), that balance architecture and human energy systems. The flow of energy through each pattern enters into resonance with the organ or function it represents, essentially "retuning" it. Karim's biosignatures can be imprinted in water for a unique way of incorporating their powerful influence.

Other symbols, such as those used by Reiki practitioners, and geometric patterns, such as the Flower of Life, the Sri Yantra, and many mandalas, carry powerful energetic significance. They too can be imprinted to water.

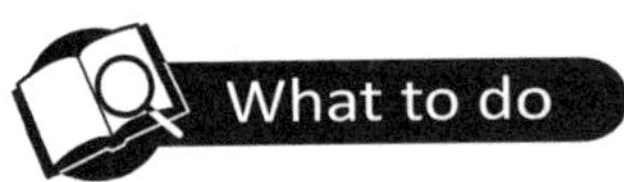

To imprint water with words or symbols:

Place a word or symbol underneath or alongside a container of structured water. Use any of the methods outlined previously to set the imprint—or simply leave the symbol until the water is consumed.

Adding information from thoughts and emotions

According to research, intention manifests as an electric and magnetic energy producing an ordered flux of photons capable of changing the molecular structure of matter.[25] The human mind continuously generates multidimensional vibrating patterns that resonate with similar vibrations. Repetitive thoughts and focused attention eventually attract their physical counterparts. Water can carry the geometric patterns of your conscious thoughts and emotions. Drinking this water holds the human crystal in continual resonance with a desired thought or emotion—without having to think about it all day long.

To imprint water with thoughts and emotions:

1. Concentrate on a specific thought and find the emotion behind that thought. Feel it from your heart as you imagine this thought/emotion entering the water. Holding the water helps but is not necessary. The magnetic potential of your thought, coupled with the intensity of your emotion, sets the imprint.

2. The program of a grateful heart is a custom with many indigenous people. Hold a glass or pitcher of water to your heart for a minute or two while feeling gratitude for the water and other blessings in your life. The electromagnetic field of your heart, along with your conscious intention, sets the imprint. Drinking this water helps you hold the spirit of gratitude, which is so pivotal for the manifestation of desires.

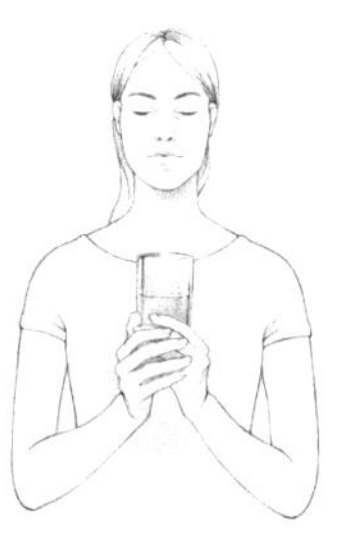

Focused attention on a larger scale

For years, Dr. Emoto advocated the use of prayer and focused attention to help heal large bodies of water. He gathered countless groups to pray for polluted water—with remarkable results. His method of testing the water before and after prayer revealed considerable differences in the water's ability to form crystals.

In 2008, Lynne McTaggart and Dr. Konstantin Korotkov collaborated with over 700 participants from all over the world to send attention to a vial of water in Russia. Dr. Korotkov's measurements of the light emitted from the water demonstrated significant changes during the time participants focused on the water *glowing*. The control vial just 2 meters away showed no significant changes.[26] This experiment is another demonstration of how focused attention can change water.

Focused attention is the creative power that forms and renews life. When it originates from the organized crystalline matrix of the human body it has tremendous power. As more people realize this, the potential to heal the water on the entire planet becomes a real possibility. The authors of *Dancing with Water* post a quarterly guided *Meditation for the Water*, which is available on the dancingwithwater.com website. It is an opportunity for the global community to gather with the intention to heal the Earth's water—and of course, it is healing for all those who choose to participate. Everyone is encouraged to take part.

Protecting water's information

Once water has received high-frequency information, weak electromagnetic fields inherent in water's vibratory matrix help to protect the information. However, water is so sensitive that it will continue to pick up energetic information from the environment. As long as the environment resonates at a high frequency, the information in water is preserved. However, if the vibratory influences from the environment are dense or of low frequency, the water has a more difficult time holding the patterns it has received. Exposure to electromagnetic frequencies, fluorescent lights, microwaves, direct sunlight, plastic containers, and even to highly charged negative emotions will eventually disrupt the structure and the information in water.

There are many ways to help shield imprinted water from outside influences. Water that is structured with seawater is very resilient and will maintain programs extremely well. Water that is rich in hydrogen and ormus is also more stable. Your own conscious intention to preserve the programming in water will protect it to a certain extent too. Additionally, you can protect imprinted water by surrounding it with a Tensor Ring, or by placing a triskelion on the water container, or by placing a quartz crystal inside. Laminar crystal, ANCHI Crystals, and paramagnetic sand also establish protective electromagnetic fields. Any of these methods will help to preserve the information in the water you create.

References (Chapter 14)

1 Smith, C. (1998). Water and Bio-Communication in *High Dilution Effects on Cells and Integrated Systems: Proceedings of the International School of Biophysics*, Casamicciola, Napoli, Italy. Edited by Cloe Taddei-Ferretti and Paolo Marotta. World Scientific, p. 295.

2 Germanov, E., Voeikov, V., Novikov, S., and Surinov, B. (2012). The Perspectives of Practical Application of Electronic Transmission of Medical Drugs: The Role of Water. A presentation delivered at the Seventh Annual Conference on the Physics, Chemistry and Biology of Water. West Dover, VT, October 19, 2012. Available online: http://www.waterjournal.org/

3 Smith, C. (2004). Quanta and Coherence Effects in Water and Living Systems. *Journal of Alternative and Complementary Medicine*, 10(1), pp. 69-78. Available online: http://citeseerx.ist.psu.edu/viewdoc/download?doi=10.1.1.605.3761&rep=rep1&type=pdf

4 Smith, C. (1998). Biological Systems in Water in *High Dilution Effects on Cells and Integrated Systems: Proceedings of the International School of Biophysics*, Casamicciola, Napoli, Italy. Edited by Cloe Taddei-Ferretti and Paolo Marotta. World Scientific, p. 93.

5 Davidson, D. (1997). *Shape Power.* Section 6.3.3. Available online: http://www.free-energy-info.com/Davidson.pdf

6 Smith, C. (1994). Electromagnetic and Magnetic Vector Potential Bio-information and Water in *Ultra High Dilution*. Edited by P.C. Endler and J. Schulte. Kluwer Academic Publishers, p. 187.

7 Smith, C. (2004). Quanta and Coherence Effects in Water and Living Systems. *Journal of Alternative and Complementary Medicine*, 10(1), pp. 69-78. Available online: http://citeseerx.ist.psu.edu/viewdoc/download?doi=10.1.1.605.3761&rep=rep1&type=pdf

8 Anderson, S. with Spurling, S. (2012). *In the Mind of a Master.* iUniverse, p. 48.

9 Smith, C. (2004). Quanta and Coherence Effects in Water and Living Systems. *Journal of Alternative and Complementary Medicine*, 10(1), pp. 69-78.

10 Personal conversation with Hans Becker 7/23/2009.

11 Germanov, E., Voeikov, V., Novikov, S., and Surinov, B. (2012). The Perspectives of Practical Application of Electronic Transmission of Medical Drugs: The Role of Water. A presentation delivered at the Seventh Annual Conference on the Physics, Chemistry and Biology of Water. West Dover, VT, October 19, 2012. Available online: http://www.waterjournal.org/volume-5/supplement-volume-5#germanov

12 Swanson, C. (2008). Effect of Pomé/Anchi Crystals on Raman Spectrum of Water. Synchronized Energy Institute.

13 Personal conversation with Glynda Yoder, cofounder ANCHI Crystals, 2014.

14 Davidson D. (1997). *Shape Power*. Available online: http://www.free-energy-info.com/Davidson.pdf

15 Simmons, R. and Ahsian, N. (2005). *The Book of Stones*. Heaven and Earth Publishing.

16 Melody (1995). *Love Is in the Earth: A Kaleidoscope of Crystals*. Earth Love Publishing.

17 Katz, M. (2004). *Wisdom of the Gemstone Guardians*. Natural Healing Press.

18 Katz, M. (2002). *Aquamarine Water: Fountain of Youthful Vitality.* Gemisphere, pp. 6-11.

19 Martino, R. (2014). *Shungite: Protection, Healing and Detoxification.* Originally published in French 2011. Translated by Jack Cain. Healing Arts Press.

20 Niwa, Y., Iizawa, O., Ishimoto, K., Jiang, X., and Kanoh, T. (1993). Electromagnetic Wave Emitting Products and "Kikoh" Potentiate Human Leukocyte Functions. *International Journal of Biometeorology*, 37(3), pp. 133-138.

21 Emoto, M. (1999). *Messages from Water.* IHM Press.

22 Davidson, D. (1997). *Shape Power.* Section 1.1.1. Available online: http://www.free-energy-info.com/Davidson.pdf

23 Emoto, M. (1999). *Messages from Water.* IHM Press.

24 Gerrard, D. (1991). *The Paper Doctor: A Vibrational Medicine Cabinet.* The Bookworks.

25 Bonilla E. (2008). Evidence About the Power of Intention. *Investigación Clínica* 49(4), pp. 595-615 (in Spanish). Abstract translated online: http://www.ncbi.nlm.nih.gov/pubmed/19245175

26 McTaggart, L. (2008). The Intention Experiment Website: http://www.theintentionexperiment.com/the-second-korotkov-water-experiment-january-18-2008

When we have thoughts and feelings of gratitude and love, water is much more responsive to us, and we are more responsive to water because we respect the life within it. This is a deeper level of the "dance with water" we are still learning.

MJ Pangman and Melanie Evans

Chapter 15

Therapeutic Water

Hydrogen-rich, Oxygen-rich, Magnesium Bicarbonate, Ormus and Deuterium-depleted

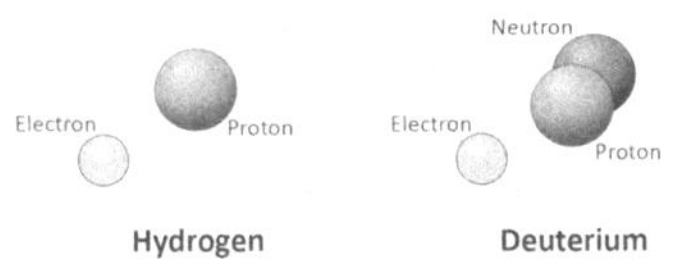

Structured water itself is therapeutic; however, there are several specific types of water that have consistent therapeutic effects.

Hydrogen-rich water is deeply cleansing. It helps balance an overabundance of free radicals while providing energy at the cellular level. The consumption of hydrogen-rich water can positively affect a broad array of metabolic functions—with resulting remediation of many symptoms (see Chapter 5). The instructions in this chapter reveal several ways to make hydrogen-rich water.

Oxygen-rich water is also cleansing. It is extremely helpful for supporting the immune system and for neutralizing pathogens throughout the body. Like hydrogen, oxygen plays a key role in many metabolic functions, including the synthesis of ATP. Regular consumption of oxygen-rich water, supports healthy metabolism and detoxification. Instructions in this chapter include ways for enriching your water with oxygen and for making ozonated water—the most potent form of oxygen-rich water.

Bicarbonate water supports the bicarbonate buffering system, which governs the body's pH; it supplies hydrogen and provides a way for minerals to be carried in organic complexes. Magnesium bicarbonate water provides magnesium in an easily assimilated form with benefits for a wide variety of magnesium-related symptoms (see Chapter 5). Following the instructions in this chapter, you can make your own.

Ormus-rich water has been of interest to those seeking heightened states of awareness since the 1990s when David Hudson's work with "white powder gold" brought renewed interest in alchemy. His work led to the exploration of a variety of methods for making ormus-rich water. According to Chris Emmons, a pharmacist and author of the book *ORMUS: Modern Day Alchemy*, the presence of ormus in the body strengthens the mind. It may contribute to coherence in the brain, which supports clarity of thought and purpose. With regular consumption, it may also contribute to a more centered feeling of well-being and an enhanced ability to deal with stress.[1,2] A variety of ways to enhance ormus in your water are presented in this chapter.

Deuterium-depleted water (DDW) has reduced levels of the hydrogen isotope known as deuterium. All water has deuterium in it—about 150 ppm. That's not much until you consider how much water is in the human body. Deuterium contains a proton and a neutron in the nucleus of the atom as shown below. When deuterium

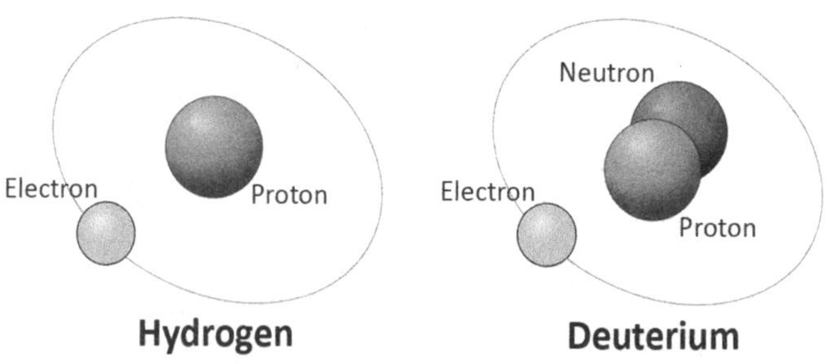

(D) combines with oxygen, the resulting water is referred to as "heavy" water, or D_2O. Because of the mass difference and the fact that deuterium bonds are stronger than hydrogen bonds, deuterium behaves differently from hydrogen in metabolic reactions. Research (mostly in Eastern Europe and Russia) has shown that deuterium slows DNA replication. It causes errors in transcription and hinders DNA repair. Deuterium also stiffens proteins and interferes with proton movement. Its presence causes the ATP mechanism to "stutter." As you might suspect, deuterium is linked with aging.[3] Many years of research have shown that drinking DDW reduces the amount of deuterium in the body, with positive effects on diabetes, tumors, many skin conditions, and numerous age-related symptoms.[4]

How to make hydrogen-rich water

The following factors are important in making hydrogen-rich water:

- **Movement**—Nature enriches water with hydrogen through movement. Vortices support water's molecular structure, which attracts hydrogen into the developing organized matrix.
- **Minerals**—The presence of minerals (especially magnesium) increases the amount of hydrogen the water will eventually hold and deliver.
- **Temperature**—Cold water holds more hydrogen (and all gases) and will enhance any method of adding hydrogen.
- **Pressure**—Placing water under pressure (in a capped bottle during the release of hydrogen) augments the amount of hydrogen held in the water.
- **Vibration**—Hydrogen is attracted to higher vibrations, so anything that raises the vibration of your water will attract more hydrogen. Love, joy, and gratitude are all hydrogen magnets!

- **Water's structure**—Even though the presence of hydrogen naturally brings structure to water, the water will absorb and hold more hydrogen when it is structured *first.*

To enrich your water with small amounts of hydrogen, consider the use of Prills (magnesium oxide), laminar crystal, Tensor Rings, ANCHI Crystals, shungite, tourmaline, quartz crystals, egg-shaped vessels, and positive thoughts/emotions. For therapeutic levels of hydrogen, choose one or more of the following:

Hydrogen stick

Dr. Hidemitsu Hayashi, a cardiac surgeon and former director of the Water Institute in Japan, was one of the scientists who conducted early research on water ionizers. He eventually determined that the benefits of ionized water were due to the hydrogen in the water, which dissipated after ionization. With this discovery, he abandoned the use of water ionizers and developed the first hydrogen stick.[5]

One way to produce a slow release of hydrogen is to allow magnesium to interact with water. In a closed bottle, under pressure, hydrogen is forced into the structure of the water.[6]

$$Mg + 2H_2O \gg Mg(OH)_2 + H_2$$

Using a hydrogen stick as directed (there are several on the market now) will add therapeutic amounts of hydrogen to your water. Within a few minutes to several hours, depending on the product, the water is ready to drink.

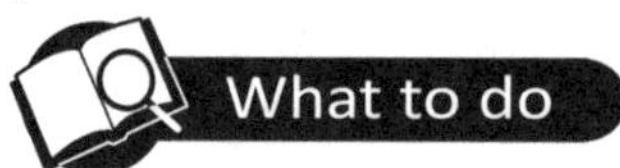

To make hydrogen-rich water using a hydrogen stick:

1. Fill a 1/2- to 1-liter swing-top bottle with structured water, or use a sturdy glass bottle with a tight-fitting lid.
2. Add the hydrogen stick and cap the bottle tightly.

3. If you like to drink cold water, place the bottle in the refrigerator (cold water holds more hydrogen).
4. After 1 to 4 hours the water is saturated, and the hydrogen-releasing reaction will come to a stop. (When a hydrogen stick is new, it will saturate the water more quickly.)

Crystal Energy

Patrick Flanagan demonstrated that a novel compound (silica hydride) provided cage-like structures within which hydrogen ions could be held.[7] He developed several hydrogen products, including a liquid concentrate called Crystal Energy, made of silica hydride and a blend of other salts. A few drops of Crystal Energy added to a pitcher of water create a matrix of silica/water cages. As the matrix forms, hydrogen is drawn in from the atmosphere. If the Crystal Energy solution is allowed to mature (several hours or overnight) it continues to bring hydrogen into the water.

To make hydrogen-rich water with Crystal Energy:

1. Add two to three drops of Crystal Energy to a 1-liter container of structured water.
2. Allow the water to mature for several hours or overnight.

Note: Because Crystal Energy creates cages to help contain the hydrogen, it is not necessary to cap the container—doing so will limit the amount of hydrogen that can be drawn into the water. The manufacturer's directions say to add 10 drops per 8-oz. glass, but if you allow the water to mature, you can use two to three drops per liter with similar results.

Active H_2

One of the easiest and most effective ways to make hydrogen-rich water is to use a product called Active H_2, an effervescent tablet made of magnesium, mannitol, and organic acids. The tablet releases hydrogen as it interacts with water. In a bottle with an airtight lid, a *supersaturated* hydrogen-rich solution can be ready in minutes with over 2 million times the number of hydrogen molecules normally found in water at room temperature.

To make hydrogen-rich water with Active H_2

1. Fill a 1/2-liter (16-oz.) swing-top bottle with structured water, or use a sturdy bottle with a tight-fitting lid. Leave about 1/2-inch of airspace at the top. If you like to drink cold water, chill the water first.
2. Place one Active H_2 tablet in the bottle and close immediately.
3. Allow the tablet to completely dissolve (at least 5 and up to 20 minutes)—longer produces the highest concentration of hydrogen in the resulting water.

Primo H_2

Primo H_2 is a unique product made of magnesium and other plant-based minerals. Although it was not designed for use directly in water, it enriches the watery environment inside the human body. Primo H_2 capsules work with digestive acids to provide a release of hydrogen for immediate absorption and uses the acidity of the stomach to sustain the reaction.

Using Primo H_2 to add hydrogen inside the body:

Take one capsule on an empty stomach up to three times a day for a gentle release of hydrogen inside the body. Primo H_2 may also be taken at night so that its regenerative effects can be utilized during rest.

How much and when to drink hydrogen-rich water

Hydrogen is cleansing—on many levels. As with structured water, adding too much, too fast, is not appropriate for most people. Add hydrogen to your regimen after your body has had time to become accustomed to drinking structured water. Start with one glass a day and pay attention to your body. Unless you are an athlete, one or two glasses of hydrogen-rich water per day is more than sufficient, and you may notice differences in energy and a variety of symptoms. For best results, drink on an empty stomach.

How to make oxygen-rich water

Oxygen-rich water is made by Nature all the time as water trips and falls over rocks and stones. In this way, water is continuously cleansed and infused with oxygen. Even shaking water will add oxygen, but not in significant amounts—and it dissipates rapidly because oxygen does not dissolve easily in water. To add therapeutic amounts of oxygen to your water, either add drops of a stabilized oxygen product (there are many different brands) or ozonate your water with a home ozonator. In the form of ozone, oxygen is extremely active and carries a greater "punch" than just oxygen-rich water. It works rapidly in the bloodstream to eradicate harmful microbes and support metabolic functions.

Oxygen-rich water made using stabilized oxygen

There are a number of concentrated and stabilized oxygen products on the market. Several drops (and up to a teaspoon in a glass of water) is enough to deliver many times the amount of oxygen normally available in water. Each product is slightly different in the way the oxygen is stabilized and delivered. Find one you feel comfortable with and follow the directions. Many companies that sell stabilized oxygen products also sell finger pulse oximeters, which validate the increase in oxygen in the bloodstream with the consumption of oxygen-rich water.

Ozonated water

In the authors' opinion, no home is complete without an ozonator. Many home units can now be purchased inexpensively. Their usefulness is worth the small investment—for water purification; production of therapeutic water for drinking, rinsing sinuses, eyes, and ears, and bathing; and cleansing meats, vegetables, countertops, and dishcloths. The release of a small amount of ozone into the air provides disinfection in the home environment too, but be careful not to inhale concentrated ozone directly because it can cause damage to lung tissue. Most home units do not release enough ozone to be harmful.

Ozonated water is the premier way to add oxygen to water and an easy way to bump up the level of oxygen in the bloodstream for a variety of therapeutic benefits. It can support many kinds of healing and is especially helpful for respiratory conditions, autoimmune problems, parasite infections, skin problems, and chronic infections. When made with the addition of seawater, ozonated water supports the antioxidant buffering system in the body (see page 58) with the therapeutic properties of both oxygen *and* hydrogen.

To make ozonated water:

1. Add 1/4 to 1 tsp. of microfiltered seawater to 10 oz. of water in a glass or ceramic container. **Note:** Never use plastic with ozonated water.
2. Ozonate for 15 to 20 minutes, depending on the instructions with your ozonator.
3. Allow the water to sit in stillness for two to three minutes (ozone creates a beautifully structured molecular array).
4. Drink within 10 minutes, one to three times a day.

Other uses for ozonated water:

- For a regenerative bath, fill the tub. Add up to 1 cup of any or all of the following salts: Epsom salts, baking soda, or sea salt. Ozonate the water for 30 minutes. Soak for 20 to 30 minutes.
- Soak fresh vegetables (also meat, fish, and poultry) in ozonated water for three to five minutes to neutralize pesticides, kill bacteria, eradicate surface contaminants, and prolong shelf life.
- Make ozonated saline solutions to rinse sinuses and eyes and to combat ear infections. To make a saline solution, add 2 to 3 tsp. of microfiltered seawater to each cup of filtered water. Add 1/8 tsp. of baking soda to buffer the solution and ozonate for 15 to 20 minutes. Gently warm and use in a neti pot for sinuses. Use a dropper bottle for tired eyes and six to eight drops in the ear canal to help ear infections.

How much and when to drink oxygen-rich water

One to three glasses of oxygen-rich ozonated water a day can be very beneficial as long as you build up to that amount. Begin by drinking one glass per day. Like other therapeutic waters, oxygen-rich water is best when consumed on an empty stomach. Take at the onset of a cold or flu to help kill bacteria, viruses, or other harmful microbes. Oxygen-rich/ozonated water will not harm healthy microflora in the intestinal track, as long as it is not consumed excessively.

How to make magnesium bicarbonate water

Magnesium bicarbonate is a salt that exists only in water. (You cannot buy magnesium bicarbonate like you can buy sodium bicarbonate.) Dr. Russell Beckett devised a way to make magnesium bicarbonate water that results from the combination of carbon dioxide and magnesium. His product, called Unique Water, is available in Australia. There are other products on the market, including

MagBicarb Water (a concentrate) and natural springwater sold under the name Noah's Water, but you can make your own with carbonated water and magnesium hydroxide (milk of magnesia).

The easiest way to make magnesium bicarbonate water is to make a concentrate—then dilute it for consumption. The authors have found that one of the best ways to maximize magnesium and balance bicarbonates in the final product is to begin with structured water made using unprocessed salts. Natural salts contain other bicarbonates (such as sodium bicarbonate, potassium bicarbonate, and calcium bicarbonate) and chlorides to balance the water. To make magnesium bicarbonate water, you will need:

- 1 liter of chilled carbonated water (also called seltzer water, any brand). Make sure it is just carbonated water with no flavorings or sweeteners. Don't use club soda because it has added sodium.
- 3 tbsp. (45 ml) milk of magnesia (The "active" ingredient should be only magnesium hydroxide—$Mg(OH)_2$. The "inactive" ingredient should be only purified water. One tablespoon should have 500 mg of magnesium. Generic brands are typically the ones without sweeteners and other additives.
- Structured water with a low mineral content (<50 ppm TDS) is best because too many minerals tend to compete with the magnesium and CO_2. This is one time when the use of distilled water is okay. However, be sure to add a small amount of minerals using unprocessed salts according to instructions in Chapter 13. The authors typically make this solution several days in advance so the water has a chance to assimilate the salts before making magnesium bicarbonate water.

To make magnesium bicarbonate water:

1. Shake the bottle and then measure 3 tbsp. of milk of magnesia.
2. Without agitating it, open the chilled liter bottle of carbonated water. Slowly add the milk of magnesia and tightly recap the bottle to minimize loss of CO_2.
3. Shake and refrigerate. The plastic bottle will become pressurized as the reaction proceeds—then as the reaction nears completion, the sides of the bottle will cave in: $Mg(OH)_2 + 2CO_2$ » $Mg(HCO_3)_2$.
4. When the liquid clears (typically within 20 to 30 minutes) the magnesium hydroxide in the milk of magnesia will have reacted with the CO_2 to become concentrated magnesium bicarbonate water with approximately 1500 mg of magnesium and 7500 mg of bicarbonate. Depending on the brand of carbonated water and the amount of CO_2 that escapes during the addition of the milk of magnesia, there may be a small amount of unreacted magnesium in the bottom of the container. It will not hinder the efficacy of the final solution and can be left in the bottom of the bottle.
5. To make a gallon of dilute magnesium bicarbonate water for final consumption, transfer 1/3 liter of the magnesium bicarbonate concentrate (333 ml) to a 1-gal. container and fill the container with low mineral–content structured water (see ingredients, previous page). The final solution contains about 120 mg/l of magnesium and 600 mg/l of bicarbonates. Depending on the source of water used for dilution, the pH will end up being between 8.0 and 8.5. (Also, see alternate step 5 on the next page.)
6. Mature overnight in the refrigerator and keep refrigerated.

Alternate Step 5

If you want to make a less concentrated, more bioavailable magnesium bicarbonate water and you are willing to wait several days, add 1 oz. (2 tbsp.) of the concentrate to 1/2 gal. of structured water. Place this in either the Water Cradle with orgonite base or an Earth resonance device for several days to mature. Both these devices refine water's structure. Magnesium bicarbonate water produced by this alternate method is more easily assimilated, so less is needed to provide the same amount of absorbable magnesium and bicarbonate. *The authors are not kidding when they say that less is necessary to produce the same results when using this alternate method.*

Water Cradle with orgonite base

Earth resonance device (see page 217)

How much and when to drink magnesium bicarbonate water

Dr. Beckett recommends two levels of consumption—a therapeutic and a maintenance level. The therapeutic dose is 2 liters per day on an empty stomach at least 45 minutes before meals. The maintenance dose is 1/2 liter per day. Most people should build up to the therapeutic dose—taking two or more weeks to get to the 2-liter-per-day level. Start with 1/2 liter per day. Drink it in two doses, one first thing in the morning and the other later in the day 45 minutes before a meal or before bed—always on an empty stomach. If you drink magnesium bicarbonate water with food, the bicarbonates will be neutralized by stomach acid produced when food is present.

Depending on your age and health, take the therapeutic dose for six months to a year, then reduce to the maintenance level. You may decide to take the therapeutic dose for two to three weeks out of each year thereafter. Pay attention to your body and adjust as necesary.

How to make ormus water

Numerous websites are devoted to the alchemy of making concentrated ormus. One of the best websites is Barry Carter's: http://www.subtleenergies.com/. One of the best texts is Chris Emmons's *ORMUS: Modern Day Alchemy*. The methods outlined here simply *awaken* the ormus elements already present in water; they are not intended to concentrate ormus. If you want a concentrated ormus product, many are available on the Internet. The addition of a concentrated product with any of the following will enhance and invigorate the concentrate.

Making ormus water using a magnetic trap

Any process or device that spins water in the presence of magnetic fields awakens the ormus elements. An easy way to spin the ormus elements into their high-spin state is to use a vortexing device that incorporates magnetic fields, referred to as a *magnetic trap*. As the water spins, oxygen is pulled downward, and hydrogen and other ormus elements are drawn upward as they enter their high-spin state. The longer the water spins, the higher the energy quotient. However, most water contains few ormus elements, so the addition of ormus-rich salts is a good idea to create more ormus in the vortexed water.

Seawater and sea salt are the starting materials for many concentrated ormus products because they contain a balanced blend of all the elements. Some salts (particularly bamboo salt and Dead Sea salt) are rich in ormus elements. These (and other) salts may be added to water prior to vortexing.

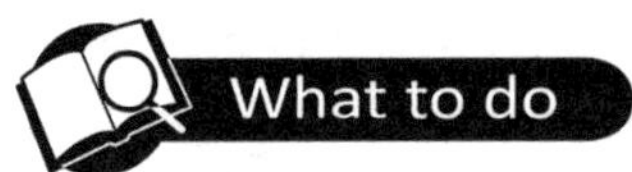

To make ormus in a magnetic trap:

1. Add ormus-rich salts (in solution) or an ormus concentrate to water.
2. Use an automated device, such as the Tribest Duet Water Revitalizer, or you can spin the water using handheld methods, such as the Vortex Magnetizer or waterfall-type devices.
3. Keep the water in motion for at least 5 to 10 minutes. Drink it immediately.
4. You can store ormus made this way for a short period of time if you keep it tightly sealed and away from light, heat, and electromagnetic fields. Even the EMFs from a refrigerator will deenergize ormus, so *do not* refrigerate it. Ormus can be stored for longer periods of time in Miron violet glass or protected with a Tensor Ring, triskelions, orgonite, or other protective methods mentioned at the end of Chapter 12.

Ormus-rich water made in an Earth resonance device

Another way to activate ormus within a magnetic field is more subtle and takes longer than the other processes but has a more profound and long-lasting effect on the resulting ormus. Barry Carter and Dr. Philip Callahan determined that paramagnetic soils are abundantly charged with ormus. Paramagnetic sand awakens ormus. It also amplifies the resonant frequencies of the Earth and the infrared wavelengths that bring structure to water.

The method outlined below is similar to a method first described by Dr. John Milewsky to create what was later called MEOW water—magnetite-enhanced ormus water.[8] Although Dr. Milewsky and others use magnetite sand, the authors have found that high-quality paramagnetic sand (with a μCGS rating of 8000 or above) is gentler and works even better. Both materials can be mixed (six parts paramagnetic sand to one part magnetite) for a nice balance between the two. The authors have created a blend they call the Earth resonance blend, which also includes quartz sand, garnet sand, and mica for enhancement. Earth resonance devices make amazing ormus-enhanced water!

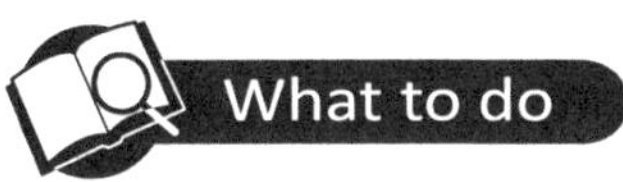

To make ormus in an Earth resonance (ER) device:

1. Make an Earth resonance (ER) device, like the one pictured, with at least 3/4 in. of paramagnetic sand surrounding the water vessel. Cardboard or glass enclosures are better than plastic or metal. (Plastic interferes with water's electromagnetic field, and metal dampens the effects of the paramagnetic material.)
2. Add ormus-rich salts (as a saturated solution) or add a concentrated ormus product to the water.
3. Place water in a glass bottle inside the device. Cap with a stone, a cork, or other natural material rather than a plastic or metal lid.
4. Allow the water to sit in a quiet place (preferably in the dark) for several days. Five days completes a cycle as discovered by John Milewsky, that produces ormus with added qualities.[9] Ormus-rich water can be kept in an ER device until consumed or placed in another suitable container and kept in the dark (but not in the refrigerator) in a protected environment until consumed.

Ormus-rich water with laminar crystal, ANCHI Crystals, and Tensor Rings

- **Laminar crystal** contains many elements in the ormus state. When it is placed in water, ormus elements are awakened, and hydrogen in the ormus state is drawn into the water from the atmosphere. Laminar crystal objects can be used by themselves or in conjunction with an ER device to awaken ormus.
- **ANCHI Crystals** also awaken ormus elements—even more rapidly than laminar crystal. They can be used by themselves or sprinkled on the top of the paramagnetic blend in an ER device.
- **Tensor Rings** have a paramagnetic value of about 18,000 μCGS, as determined by Phil Callahan.[10] They can be used by themselves or with an ER device to awaken ormus.

To make ormus using laminar crystal, ANCHI Crystals, Tensor Rings, and Earth resonance:

1. Add ormus-rich salts (in a saturated solution) or add a concentrated ormus product to water. Place laminar crystal objects (sets of three are best) directly in the water. Allow the water to sit uncovered for 6 to 12 hours, depending on the size and number of laminar crystal objects you use. Alternatively, you can place the container with laminar objects inside an ER device for several days (see diagram opposite page).
2. Add ormus-rich salts to water, as in step 1. Then place a pouch of ANCHI Crystals around (in close proximity to) the water. Allow the water to sit for several hours or overnight. Alternatively, you can place the water inside an ER device and sprinkle ANCHI Crystals on top of the paramagnetic blend (see diagram). Leave for several days.

3. Add ormus-rich salts to water, as in step 1. Then place a Tensor Ring around the container or pour water through the Tensor Ring several times. Allow the water to sit for several hours or overnight. Alternatively, you can place the water container in an ER device with a Tensor Ring around it (see diagram). Leave for several days.
4. All of the above.

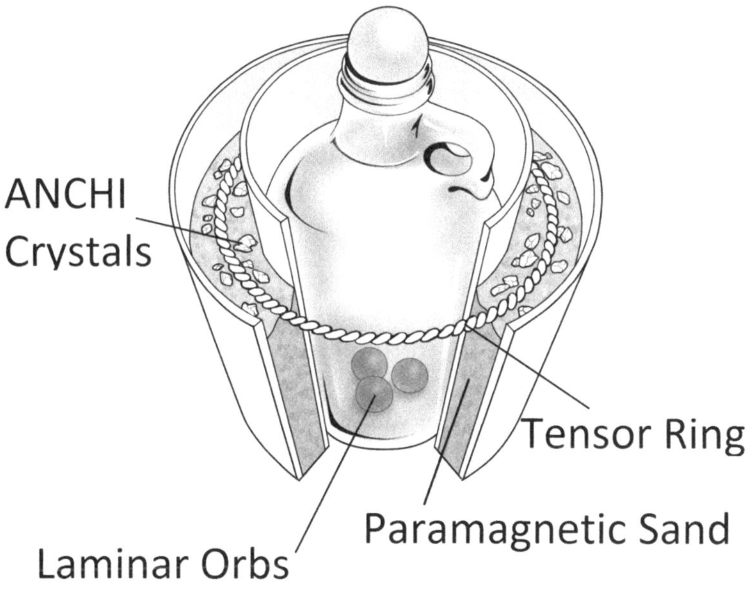

An Earth resonance device incorporating ANCHI Crystals, laminar orbs, and a Tensor Ring

How much and when to drink ormus-rich water

People drink ormus-rich water for a variety of reasons. Some like it on an occasional basis to enhance meditation or ceremony. Others drink it on a more regular basis to experience a wide range of benefits that include greater mental clarity, heightened awareness, an enhanced sense of well-being, and an ability to deal with stress. Still others use it in the garden, because numerous trials have shown plants respond favorably to ormus in the water. Whatever your purpose, ormus is best taken on an empty stomach, when these highly energized elements can disperse rapidly into the parts of the body where they are needed.

Deuterium-depleted Water

Nature provides deuterium-depleted water (DDW) at high elevations, because water with the greatest concentration of deuterium (heavier) precipitates out first, leaving deuterium-depleted water to fall as snow. Mountainous and polar regions of the world have water with a lower deuterium content. (Water at the equator is typically 156 ppm, and water in the Rocky Mountains, at an elevation of 5700 ft. (USA), has a deuterium content measuring 136 ppm.[11]) Current technologies for producing DDW are costly, making it difficult to provide a high-quality product for consumption. In the United States at the time of this writing, there is no one producing DDW with a significantly lower deuterium concentration. However, a 25-ppm product called Qlarivia is made in Romania and can be purchased online.

How much and when to drink DDW

Those who advocate drinking DDW recommend 1 liter per day (100 ppm deuterium or less) for one to two months every year. A smaller amount is required if a lower concentration of DDW is available. For long-term benefits, drink DDW every day. For further information, see Appendix C.

There are many "designer" waters being developed. Some of these products have been amended with vitamins and herbs. Others have been imprinted with specific *information.* Although these products can be therapeutic, some may carry information that is not necessarily biocompatible or beneficial for *everyone.* In deciding whether to purchase *designer* water, be sure you understand the ingredients or frequencies that are incorporated and whether they are beneficial for you. Making your own therapeutic water (designed specifically for you) using the guidelines in this book may be in your best interest—it will certainly be more fun. When you incorporate therapeutic water, proceed slowly and pay attention to subtle changes in your body—cut back if you notice fatigue or other symptoms of detoxification.

References (Chapter 15)

1 Emmons, C. (2009). *ORMUS: Modern Day Alchemy*, Dreamgate Press.

2 Emmons, C. YouTube video: https://www.youtube.com/watch?v=1ZYFhC72dgQ

3 Silva, D., Somlyai, G., Somlyai, I., and Aschner, M. (2012). Anti-aging Effects of Deuterium Depletion on Mn-induced Toxicity in a *C. elegans* Model. *Toxicology Letters*, 211, pp. 319–324. Available online: http://deuteriumdepletion.co.uk/download/DDW_anti_aging.pdf

4 Dancing with Water Website: http://www.dancingwithwater.com/deuterium-depleted-water-what-is-it/

5 Hayashi, H. *Hydrogen-Rich Water Guidebook*. Chapter 8. Available online: http://www.new-water.org/world/index.html

6 Hayashi, H. *Hydrogen-Rich Water Guidebook*. Chapter 8. Available online: http://www.new-water.org/world/index.html

7 Stephanson, C. and Flanagan P. (2003). Synthesis of a Novel Anionic Hydride Organosiloxane Presenting Biochemical Properties. *International Journal of Hydrogen Energy*, 28(1), pp. 1243-1250.

8 Milewsky, J. (2003). The Effects of Magnetic Water. Available online: http://www.subtleenergies.com/ormus/tw/magneticwater.htm

9 Milewsky, J. Subtle Energies Website. Reported in an online article: http://www.subtleenergies.com/ormus/tw/magneticwater.htm

10 Garrison, C. (2004). *Slim Spurling's Universe*, IX-EL Publishing, pp. 68-69.

11 Analytical report from the National Research and Development Institute for Cryogenics and Isotopic Technologies, Valcea, Romania. In the possession of MJ Pangman.

Chapter 16

Treating the Water for Your Whole Home and Garden

The water you bathe or shower in has an impact on your health too—because water is absorbed through the skin. Similarly, the water used in your garden can influence growth, drought tolerance, insect resistance, and the nutritive value of plants. Methods for structuring water as it enters your home are based on the same principles used with smaller batches for personal consumption. And as with smaller batches, the water should be free of contaminants.

Contaminants

Any method that brings life force to water changes the nature of the water. If water is able to sustain a high vibration for a long enough period of time, many harmful microorganisms will die or go elsewhere; they cannot thrive in a sustained higher resonance.[1] Under some circumstances, water-structuring devices have been shown to render contaminants less harmful (neutral in their effects on biological systems), but the degree to which this can be accomplished as water passes through the pipes in your home depends on a number of factors, including the nature of the contaminants and how heavily the water is burdened. Although the manufacturers of some water-structuring devices make claims about neutralizing contaminants, these devices should not be counted on to deliver clean water. Especially if your water is heavily burdened, it is important to implement some form of filtration or purification *before* you structure the water that enters your home (see Chapter 11). Just as with smaller batches of water, when the burden of pollutants is released, water is open to receive life force, which can be gathered in many ways.

Hard water

Structuring devices go a long way toward eliminating the need for traditional water softening (see page 22). Structured water makes your skin and hair feel softer without the slick or slippery feeling of traditionally softened water. Although structuring reduces scale, it does not eliminate the minerals that cause it. Water spots on dishes and fixtures can still be a problem if water is allowed to dry on them. On the other hand, these deposits are much easier to remove because the minerals don't form traditional scale deposits; a little vinegar will typically remove them. Structuring devices lengthen the life of hot water heaters, dishwashers, air conditioners, evaporative coolers, and other equipment—and they will not harm the environment. If anything, water that is structured supports the environment. It contains more energy as it leaves your home and may also deter some forms of algae and bacteria.

Returning life force to the water in your whole home

Most people are surprised by how quickly water responds to spiral movement. Many devices for the whole home incorporate spiral flow forms. Some are connected *inline*, and others are fastened to a water outlet, such as a faucet or shower.

The use of magnets is another way to bring structure to the incoming water. They may be used as a stand-alone system, but they are easily combined with other methods for a more potent effect. Water passing through a magnetic field is affected by the configuration of the magnets and the rate of flow. The authors especially like the effect of paramagnetic forces on water. Paramagnetic sand and Tensor Rings (also paramagnetic) create an invigorating energetic environment that carries Earth resonance and many harmonics from the mineral kingdom to balance and structure water. Both can be implemented on incoming water lines or at exit points.

Some water-structuring devices use resonance and/or geometry to accumulate and communicate life force to passing water. Because their impact is energetic rather than physical, they may *seem* less able to structure the water, but their influence should not be underestimated. Some of these devices are well supported by research. Below are examples of each of the above types of devices with websites for you to do further research. The list is by no means all inclusive.

Water structuring based on spiral movement

Alive Water Systems (Canada). These devices are very simple; an internal copper spiral provides spinning motion for the water. Devices include inline models and those that attach to the kitchen sink or shower outlet. They are made of either copper or stainless steel.
http://alivewater.com

Fractal Water Technology (USA). Fractal Water's imploder technology is based on spiral movement and golden ratio proportion. These devices spin water in one direction and then the other. The Super Imploder combines high-density spin with magnets, creating

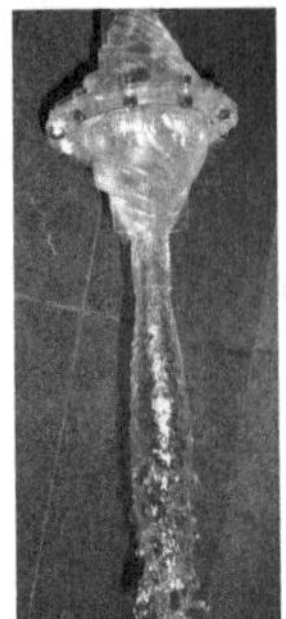

a powerful water-structuring device available for home or garden. (Pictured are the shower model with no magnets and the Super Imploder.) They are made of Fortran, a resilient plastic.
http://fractalwater.com

Natural Action Technologies, Crystal Blue Enterprises, Greenfield Naturals, Rainmaker, and President Water (USA). These product lines are all spin-offs on the same concept: spherical flow forms are positioned inside a tube that is plumbed inline. They are made of a variety of materials, from plastic and glass to stainless steel. Some include resonance technologies that may not be biocompatible for everyone.
http://naturalactiontechnologies.com/
http://www.crystalblueent.com/
http://greenfieldnaturals.com/
http://rainmakerh20.com/
http://presidentwater.com/

Water Whirls (Germany). Water Whirls, created by Carat Aqua Plus, guide water through two inlet channels into a vortex chamber where water is sucked upward and then downward, spinning in the opposite direction. Using centripetal and centrifugal forces, the resulting water is energized and oxygenated as it exits the faucet. They are beautifully crafted of titanium with a standard thread pattern that fits any kitchen faucet. The company also makes a line of exceptional showerheads.
http://www.carat-aquaplus.com/

Artesia fountains (USA). Randy Hatton makes these distinctive vortexing fountains that are functional as well as beautiful. Each piece is a work of art, designed uniquely for pools, spas, Jacuzzis, and fish ponds and is based on golden ratio vortex motion with energetic enhancements.
http://vibrantvitalwater.com

Water structuring based on magnetics/electromagnetics

There are many magnetic water conditioners—everything from permanent magnets to electromagnetic devices that attach to water lines. Some people like to make their own magnetic array, and as long as the magnets create a strong repelling force, they will have an impact on water. Strong magnets are best (especially if you have metal pipes), but be very careful when you work with strong magnets; they can pinch fingers.

Water structuring-devices based on resonance and/or geometry

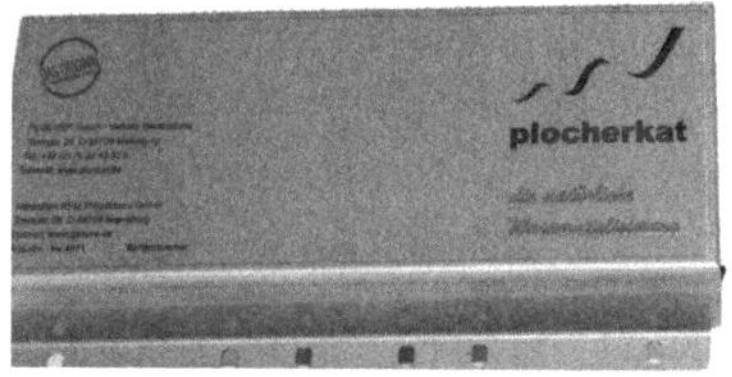

Plocher (Germany). Plocher devices bring structure to water via resonance. A wide variety of Plocher systems are imprinted using energy accumulators and imprinting technology inspired by Wilhelm Reich. Plocher has also developed a variety of energized additives that have been extremely successful for reinvigorating small bodies of water and agricultural land. These devices are made of stainless steel and do not require plumbing; their effects are permanent and maintenance free.
http://www.plocher.com/
http://www.symbionature-usa.com/ (in the USA and Canada)

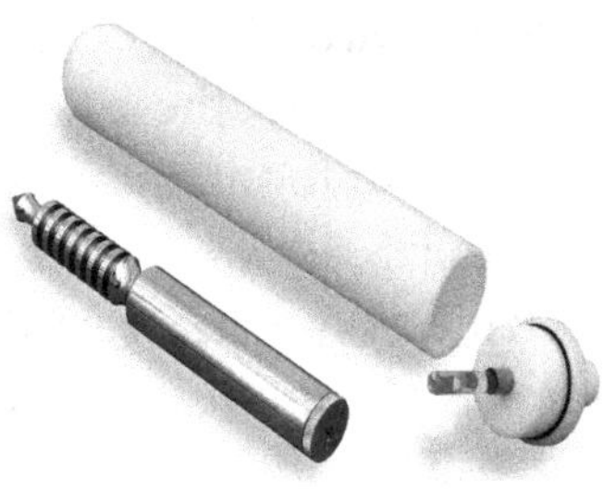

Weber-Isis (Germany). Weber-Isis devices are based on the geometry and resonance of the Egyptian *djed*, which is a powerful accumulator of life force. The concentric rings on the djed are what Water Russell would call *loops of force*.[2] They concentrate and amplify life force. These devices are made of gold-plated brass with either plastic or copper enclosures. Some of the larger devices are enhanced with quartz sand. Attached to the water line, no plumbing is required. The effects are permanent and maintenance free.
http://weberbio.de/weber-isis-wasseraktivator-mobil.html?store=english&from_store=default

Grander (Austria). The Grander water systems, developed by Johann Grander, send water through a chamber adjacent to "informed" water, created by Grander's proprietary method. In the authors' experience, these systems may not be biocompatible for everyone. They are made of stainless steel and require plumbing.
http://www.grander.com/

Make your own

You can make your own structuring/enhancement device using any of the concepts presented in *Dancing with Water*. These devices can often be just as potent and more natural than the more expensive devices. Some ideas follow.

Paramagnetic sand

Paramagnetic sand is an amazing gift from Mother Earth. When an adequate amount is placed around water, it causes rapid alignment and realignment of water molecules as water passes, in much the same way as vortexing water within a magnetic field. Old patterns are released, and water is allowed to gather life force. The process awakens ormus as it brings in Earth resonance and organic light.

Paramagnetic sand can be used to treat your incoming water by placing it around the incoming water line. To amplify its effects (requiring a smaller volume), quartz tubes may be used to hold the sand (or a blend of paramagnetic sand and other supporting materials) in place. The arrangement has potent, permanent effects, and no plumbing is required.

Tensor Rings

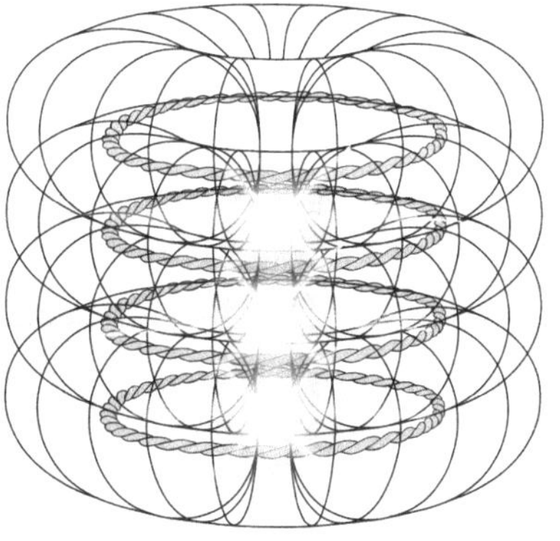

Tensor Rings create a toroidal energy field that releases vibratory pollution from water and initiates greater organization. When rings are placed in series (even numbers), they open a conduit that draws in organic light (life force) from the universal source field. The effect is greater than the sum of individual rings. Tensor Rings can be positioned on a horizontal water line so that they are between ¼ and ½ in. apart when you plumb a filter or other water treatment device. They can be used in conjunction with many other devices. Their effects are permanent and maintenance free.

Triskelions

The triskelion (made of copper, silver, or gold) is one of the simplest, most versatile devices for use with water. It adds piezoelectric and magnetic energy to water, helping to release dense vibratory pollution and raise water's overall frequency. Although less potent than paramagnetic sand or Tensor Rings for treating the water in your entire home, triskelions can be used in conjunction with any of the devices or processes mentioned in this book. You may want to place triskelions just before water exit points. This will boost the energy of the water immediately prior to use.

Crystals and other stones

Quartz amplifies any energy in its presence. It is one reason the authors like quartz tubes to amplify energetic effects. Any of Mother Earth's stones can be placed inside a quartz tube, which is then strapped to the water line to amplify the effects of the stones as water passes. This works best when it is preceded by a structuring device and the water is prepared to receive information.

Conditioning the water for your garden

The distance water travels through straight pipes after it has passed through (or alongside) a treatment device, determines the degree of structure it can retain. Many factors contribute to a loss of energy and structure, including Wi-Fi in the home, power lines, geothermal stress, and the type of pipe. Some of the devices and methods mentioned in this chapter can be used on the end of a garden hose to provide freshly structured water for the garden.

Another way to create energized water for the garden is by placing enhancements in a bucket of water. Left overnight or for 6 to 12 hours, the water can then be diluted using a fertilizer injector or siphon and applied to the ground or to foliage of plants. Use such things as crystals and stones (shungite, ANCHI Crystals, quartz, tourmaline, and others), laminar crystal, or paramagnetic sand in a crystal tube.

There are many structuring devices on the market, but not all of them produce biocompatible water. Follow your personal guidance when selecting a product, and then listen to your body as you drink the water.

References (Chapter 16)

1 Plocher Website, Plocher Guiding Principle: http://www.plocher.com/plocher-system/

2 Hayes, A. Cosmology of Walter Russell Website: http://walter-russell.com/documents/coildesign.html

Chapter 17

The Dance with Water
What to Expect

Mu Shik Jhon

In 1986, Dr. Mu Shik Jhon presented a new theory on aging. He called it *The Molecular Water Environment Theory*. According to his theory, aging is a loss of structured water from organs, tissues, and cells—accompanied by an overall decrease in total body water. He said, "It is the structure of the water within our bodies that ultimately determines health or sickness."[1] He advocated the consumption of structured water to increase vitality, slow the aging process, and prevent disease. Dr. Jhon was one of the first to document some of these benefits.[2]

Full-spectrum living water is a *whole food* and a source of *energy* for the human body. On a physical level, regular consumption may improve hydration, support more effective detoxification, enhance nutrient absorption, and improve cellular communication. On an emotional level, it may help with clarity, well-being, and an ability to deal with stress. The authors believe it can also open the doorway to heightened awareness and expanded consciousness.

Hydration

When there is insufficient water in the human body, every metabolic function suffers. Most diseases are accompanied by dehydration. This was documented by Fereydoon Batmanghelidj, MD, in his books *Your Body's Many Cries for Water* and *Water: For Health, for Healing, for Life.* According to Batmanghelidj, most so-called incurable diseases are nothing but labels given to various stages of chronic dehydration.

Ideally, water in your body should be distributed so that 60% is inside your cells. The remaining 40% should be available in the fluids outside your cells—in the blood, lymphatic tissue, and extracellular fluid that bathes every cell. For most individuals, age is accompanied by a gradual loss of water from inside the cells and an increase in water outside the cells. This shift occurs for a variety of reasons.

1. **Thirst mechanism decreases** – The thirst mechanism diminishes with age and a less active lifestyle.[3] Older people don't drink as much water. And unfortunately in our modern world, there are plenty of "nonwater" options that often take precedence.
2. **Toxin buildup** – Age-related toxin/waste buildup in the fluid around cells causes water to be drawn from *inside* the cells to neutralize toxins in the extracellular fluid. Water retention, or edema, often results, which is the body's attempt to dilute wastes that cannot be flushed from it.

3. **Lack of energy in the water** – When the energetic quality of water is low (a lack of structure, coherence, and high-vibratory information) hydration is compromised. Water receptors in our cells are "tuned" to receive highly energized, structured water. When water vibrates at lower frequencies, it is not admitted through the cell membrane. This sensitive mechanism protects cells from vibratory pollution, but it also contributes to dehydration. Most of the drinking water on the Earth today is *flat* from a vibratory perspective. It is not absorbed well. To get this water into cells, the body must revive it by drawing on its own energy reserves. When we are young, our bodies have enough extra energy to revive energetically poor water. But as we age, our energy diminishes, and it is more difficult for the body to revive the water. No wonder the thirst mechanism also declines.
4. **Free-radical buildup** – In a polluted, hydrogen-depleted world, free radicals linger. The body is forced to draw hydrogen from water inside cells. Years of drawing on cellular water to harvest hydrogen and maintain free-radical balance reduces the water inside cells.
5. **Lack of exercise** – Water needs to move to stay energized—and so do you. Physical movement calls water into cells. Without some form of organized *whole body* movement (swimming, walking, dancing, etc.), the water in your body simply follows a pathway through your organs to an exit. In this case, drinking water becomes more like irrigation than true hydration. Movement maximizes hydration. If you're not moving, and breathing deeply, your cells are not *asking* for water. Drinking *full-spectrum living water* in conjunction with whole-body movement is one of the most important keys to longevity. It is one more reason to *dance with the water.*

6. **Exposure to unnatural EM fields** – The last reason for age-related dehydration has little to do with age, but it may be one of the biggest contributors to the problem. Water does not easily enter the cell in the presence of strong electromagnetic (EM) fields like those associated with cell phones, computers, microwave ovens, electronic games, television, radar, and the alternating current that powers our homes. The human body is an electromagnetic system. Nearly every biological function depends on electromagnetic signals. Unnatural EM fields create chaos and confusion at the cellular level and hinder the absorption of water.

Many common symptoms that are considered *normal* with age are directly related to dehydration. These are often relieved with the consumption of *full-spectrum living water*.[4] However, the body is not a simple container that can be replenished when empty. When dehydration has become symptom producing, rehydration at the cellular level can be slow.[5]

Common symptoms directly related to dehydration—often reduced with the regular consumption of *full-spectrum living water*

Asthma and other respiratory problems
Blood sugar imbalances
Constipation
Fatigue
Headaches
High cholesterol
Mental fog
Skin disorders

Detoxification

When the human body is hydrated and has enough energy, it is able to cleanse itself on a regular basis. But when water lacks energy and structure, it cannot remove toxins and wastes, so they are stored in a variety of places, including in artery walls, joints, and within fatty tissue. *Full-spectrum living water* can provide a boost in cellular efficiency and energy—often enough to begin to clear stored toxins.

Your body uses the same mechanisms to remove toxins during a cold as it uses to remove stored toxins. Cleansing too rapidly can be much like having a cold or flu, and it can cause what is referred to as a *cleansing crisis*, typified by symptoms that include aches and pains, headache, fatigue, diarrhea, and/or mucus discharges. A person may also notice changes in bowel movements, urine, skin, and breath. Each of these changes can be a sign that the body is discharging toxins too rapidly. If you initiate a cleansing crisis by drinking too much structured water, cut back and drink *un*structured water until your symptoms subside—then begin again more slowly. Bathing in structured water will help if you experience a cleansing reaction (see page 241).

Proceed slowly as you incorporate structured and other therapeutic kinds of water into your diet. Give your body the time it needs to remove toxins slowly rather than all at once. Detoxification was not meant to happen overnight. It can take months—even years—to release the deeply held toxins that your body has carefully hidden to protect critical organs and systems. Heavy metals are some of the most difficult toxins to remove. Adding fulvic acid to your water may help draw heavy metals from your body and carry them away. If you know you have a problem with heavy metals, it may be best to work with a health practitioner.

Nutrient and pharmaceutical absorption

Because of its crystalline organization, *full-spectrum living water* enhances nutrient absorption. It literally *introduces* nutrients for absorption at the cell membrane with more vibrancy than typical water. *Full-spectrum living water* can introduce medications more efficiently, too, amplifying pharmaceutical signals in the same way water amplifies biological signals. So *if* you take medications, take them with unstructured water—at least a half hour before or after drinking structured water. Some people have been able to reduce medications with the ongoing consumption of *full-spectrum living water.* You may want to work with a health practitioner.

Cellular communication

Dr. Jacques Benveniste likened the role of water to the role of a disc player in playing a compact disc. He said that the disc itself does not create sound; sound is produced only when the signals recorded on the disc are accessed and amplified. The cells of the human body and their components (DNA, enzymes, etc.) are capable of accessing and amplifying the information available in water. If that information is carried clearly and coherently (if the disc is not scratched or warped), the information can easily be accessed. But if the water lacks energy or has lost coherence, then signals are dampened, and full access is denied. This was demonstrated by Dr. Benveniste and others. Although this work is still disputed, a growing body of evidence indicates water carries information and participates in cellular communication. In a paper titled *The Biophysics of Water Memory*, the researchers had this to say:

> Results suggest that electromagnetic transmission of biochemical information can be stored in the electric dipole moments of water in close analogy to the manner in which magnetic moments store information on a computer disc. The electromagnetic transmission would enable the vivo [inside the body] transmission of the specific information between two functional bio-molecules.[6]

How much to drink

Just as you cannot detoxify overnight, you cannot rehydrate overnight. Begin by drinking one or two glasses of structured water every day for several weeks, along with enough unstructured water to stay hydrated. This is sometimes enough to notice a difference in energy, bowel movements, skin, and so on. Gradually work your way to the point where you can drink as much *full-spectrum living water* as you want. Then if you wish you can begin to incorporate other water enhancements.

Although structured water often causes more frequent urination (in the beginning) it may ultimately necessitate fewer trips to the bathroom. Some individuals who make the transition to *full-spectrum living water* notice they require less water than previously. Listen to your body and follow its guidance.

What to expect

Water is as much a living entity as the vegetables you grow in your garden, and cultivating your relationship with water is as important as cultivating a garden. It requires attention and love. Water will reflect back to you every gift you bestow on it. Adding life force and other enhancements will be reflected back to you in a renewed sense of vibrancy and vigor in your life.

Those who begin to drink *full-spectrum living water* often notice an improvement in energy. Some notice softer skin. Many notice improved bowel regularity, blood sugar balance, and greater clarity. Athletes often notice they can work out harder and recover faster. On another level, some notice differences in how they feel about life, an increased ability to cope with stress, and an openness to new ideas. The possibilities are limitless when you consider the variety of ways water can be imprinted.

Informed water carries purposeful information to every cell each time you drink. However, occasionally, the information conveyed in water may go unrecognized when there is *stuck* energy in the body that needs to be released (in the form of a belief or past experience that needs to be discharged). In these cases, drinking *full-spectrum living water* may help to initiate the process. You may get ideas or be serendipitously introduced to a facilitator who can help. Water works on many levels; it is an important key to connecting with your highest self.

Supporting the liquid crystalline state of the water in your body
Although a majority of individuals who begin to drink *full-spectrum living water* notice differences in their health and well-being, some do not. We are all unique. If you feel as though you may be missing some of the benefits, consider the following:

Our modern environment produces stress, chaos, and unnatural electromagnetic fields. These disturbances are like *static* or *noise* to the liquid crystalline components of our bodies. If you carry a cell phone, if you spend long hours at a computer, if you work near electronic devices, if you have a Wi-Fi network in your home, if you sleep with electric appliances in your energy field (even if they are turned off) you are exposed to unnatural electromagnetic static. It is very difficult for structured water to upgrade the liquid crystalline matrix when it is continuously exposed to dissonance. It is like trying to stabilize a power source during an electrical storm. If you neglect to offer your body an opportunity (on a regular basis) to discharge extraneous energy, drinking the best water in the world may go unnoticed.

Many companies sell electromagnetic (EM) shielding devices. They can help, but in today's world, it is almost impossible to completely protect yourself. Everyone should consider some regular form of destressing. Following are a few suggestions to help keep the liquid crystalline water in your body structured and healthy.

- Spend time in Nature. Find places away from the chaos of the city where there is an abundance of negative ions, where there are trees and the air is fresh, where you can be exposed to the sounds of Nature.
- Be involved in regular organized movement, for example, dance, tai chi, swimming, walking, or yoga. A new book, *Cosmic Swimming* by Marion de Sirius, elaborates on the benefits of spiral symmetric swimming. This method is excellent for organizing the liquid crystalline body matrix.[7]

- Breathe. Breathing is a form of movement. For those who may be unable to exercise, deep breathing is a great way to move water into the cells of the body.
- Bathe in structured water: A good mix is 1 cup Epsom salts, 1 cup baking soda, 1 cup unprocessed salts of choice. Add essential oils, flower essences, or crystals/stones. You may also want to ozonate the water for an infusion of oxygen that can be very therapeutic. Tone, or play the sounds of Nature.
- Create sacred space in/around your home using such things as crystals, diamond water, orgonite, water fountains, or salt lamps. Many of the ways for structuring water will also affect the energy in an entire room. For example, an egg-shaped container filled with water will alter the energy of a small room. So will a good-sized piece of orgonite and/or shungite. Water (and the water in your body) responds to the energy in its environment—including the energy in your home. An enhanced environment will help to support the structure of the water in your body.
- Play music (especially music composed in 432 Hz tuning), or sounds from Nature. Sounds from the natural world will release stress and alter the structure of the water in your body.
- Wear stones and crystals. Many, like ANCHI Crystals and shungite, will help your body hold a high vibratory level and help shield you from outside dissonance. Cleanse them regularly in the sun, in diamond water, or in hot and cold running water.
- Meditate. The altered state creates organization in the body's energy field, which affects the organization of your internal water.
- Let go of anger and frustration. Negative emotions are lower vibrations and are disruptive to your body's water.

Vibrating at a new level

Water is intended to carry the life force of the universe. The Earth's natural organizing forces structure and activate water so that the codes of life can be delivered to the inhabitants of this beautiful planet. When water is allowed to dance, when it is allowed to flow within the natural channels the Earth has intelligently created for it, water delivers these codes to our DNA. *Full-spectrum living water*, made using the methods outlined in this book, has the capacity to support the physical body. It may also awaken dormant DNA codes and harmonize mind and spirit.

Your dance with *full-spectrum living water* will influence those around you. For each person who learns to dance, the challenge becomes easier for the next person . . . and the next. Water has been asking us for millennia, "*May I have this dance?*" Accepting the invitation could be the beginning of renewed life—for you and for the entire planet.

The authors have created a website that serves as an addendum to this book—a place to share new developments and further information. It includes a number of audios and videos (radio/TV interviews, speaking engagements, and product demonstrations), as well as in-depth articles on a variety of subjects not covered in the book. The website is also a gathering of the authors' favorite water-structuring tools (including some they have developed themselves) and a place to connect with the global community to participate in a quarterly guided meditation to help heal the Earth's water. The Afterword (page 245) is one such guided meditation.

www.dancingwithwater.com

References (Chapter 17)

1 Jhon, M-S. (2004). *The Water Puzzle and the Hexagonal Key*. Uplifting Press, p. xiv (preface).

2 Jhon, M-S. (2004). *The Water Puzzle and the Hexagonal Key*. Uplifting Press.

3 Editorial (1984). Thirst and Osmoregulation in the Elderly, *Lancet*, pp. 1017-1018.

4 Pangman, MJ. (2007). *Hexagonal Water: The Ultimate Solution* revised edition. Uplifting Press.

5 Bray, M., Founder, Advanced Health Clinic in Farmington, UT. Personal conversation 2016.

6 Widom, A., Srivastava, Y., and Valenzie, V. (2008). The Biophysical Basis of Water Memory. A paper presented at the XVI International Conference and Debating Scientific Club: New Information Technology in Medicine, Pharmacology, Biology and Ecology, held in Ukraine, May 31 to June 9. Available online: http://jacques-benveniste.org/bio_conf_widom.pdf

7 De Sirius, M. (2013). *Cosmic Swimming*. Books on Demand. GmbH. See also: http://cosmic-swimming.blogspot.com/

Afterword

Forest Bath: A Guided Meditation

Find a comfortable place where you can relax and be undisturbed. Take a deep breath through your nose . . . then let it out through your mouth, and as you do, let go of any tension or worries. Take another breath, filling your lungs from the bottom to the top . . . and let it go . . . softly and gently through your mouth, relaxing your entire body and mind. Consciously let go of any burdens you are carrying and allow the kind of relaxation that brings renewal and insight.

Imagine you are in a lush, green forest. The ground beneath your feet is soft with layers of leaves. Immerse yourself in the experience as you begin to follow a path. The air is damp and feels so good to breathe. Listen as you stroll among the trees. . . . Hear the birdsongs and the humming of occasional insects. Smell the pollen, the leaves, and the bark on the trees.

A soft rain begins to fall—and yet it is not really rain. The drops are so tiny and light you barely notice them as the dampness begins to collect on your skin. The moisture increases, and you become aware of a cleansing energy that has arrived with this falling mist. You feel your whole body invigorated—inside and out—as ions in the misty air completely surround you and draw all forms of negative energy from your outer body and then from deeper within. Your lungs bask in the energy that comes in with each breath of ionized forest air. They feel so fresh and clean—expanding and contracting as they deliver energized oxygen to your cells. The cleansing is deep and magical—healing—and you feel alive and at peace.

Tilting your face upward and throwing your arms outward, you feel so connected to all of life. You realize you are literally bathing in the forest . . . rich with ormus minerals and probiotic microbes. For the first time in a long time you feel at home—a part of the web of life. You belong here, and you know you were meant to feel joy in your experiences and your creations. The sense of it all is powerful. Be still. . . .

In the stillness, your ears pick up a faint roar in the distance, and the path you are on appears to lead you in its direction. You eventually come to a stream that plummets off a wall of rocks. You gaze at the falling water and the mist, so full of life force. Then you make your way to the base of the waterfall, where there is a place to sit and watch the falling water tumble over the rocks. Notice the spirals—circling left then right—and sense the buildup of energy as the water is enhanced through movement and sound.

You decide to follow the stream, walking along the bank until you come to a place where the water enters a small, protected alcove. Here, the water is still and deep. Dipping your hands into the water, it feels soft, cool, and smooth. Just holding it in your hands is invigorating.

Notice your reflection in the water and realize that you are the water and the water is you. With this realization your heart opens, pouring love and gratitude into the water . . . and at that moment you are infused with the same. Take this moment to thank the water for the life she makes possible within and around you. . . . And thank Mother Earth as you send a soft, steady flow of gratitude and healing from your heart. . . . You are aware of others all over the Earth sending love and healing at this time, and you consciously join your energy with theirs. See the love circulating through the waters of Earth, balancing and cleansing wherever it goes. . . . Notice the changes that take place because of your love and the love of others. . . . See the renewal of the entire planet. . . . Watch . . . and when you are complete, take a deep breath and let it go. Wiggle your fingers and toes . . . and open your eyes, mindful of retaining the freshness and renewal and energy of your forest bath.

To be notified of each new quarterly meditation, visit the *Dancing with Water* website:

www.dancingwithwater.com

click the tab: Heal the Earth's Water

Appendix A

Gas Discharge Visualization (GDV)
Quantifying the Photonic Energy in Living Systems and Its Applications with Water

According to the modern German scientist Dr. Fritz Albert Popp, well-known for his work with biophotonic energy: "We know today that man is essentially a being of light. We now know for example, that light can initiate, or arrest cascade-like reactions in the cells, and that genetic cellular damage can be virtually repaired, within hours, by faint beams of light. We are still on the threshold of fully understanding the complex relationship between light and life, but we can now say emphatically, that the function of our entire metabolism is dependent on light."

Gas discharge visualization (GDV) is an electrophoton capture technique designed to measure photonic (light) energy fields. The term "GDV" is shorthand for a complicated description of the process. More precisely it is known as Biological Emission and Optical Radiation Stimulated by Electromagnetic Field Amplified by Gas Discharge with Visualization. It is based on the *Kirlian effect,* which measures the light emanating from all living things. The Kirlian effect refers to the *glow* captured on photographic film appearing around the edge of a object following its placement in a high-intensity electrical field. GDV technology, invented by Dr. Konstantin Korotkov, is being used by scientists, doctors, and integrative health practitioners as a research and diagnostic tool. The combination camera, computer, and software is capable of providing data on the physical, emotional, physiological, and spiritual state of those tested.

Krishna Madappa, who works closely as a research partner with Dr. Korotkov, performed much of the work featured in the appendix. He has this to say about GDV research with water:

> For nearly a decade, our research and exploration using the Electro Photon Capture/GDV technology has shown the positive effects of "informed" or structured water, on the human energy field. Our consistent observations for those who drink this kind of water are: organization of the sympathetic and parasympathetic energy fields; organization and greater brain hemisphere balance; potentiation of chakras (energy centers) especially the Anahata (heart center); increase of bio-photons in the energy field; and improvements in other processes that are supportive to whole person wellness. We observe, in principle, that enhancement of all the above attributes infuses cascades of light which stimulate a self organizing awareness that fosters an awakening mind.

GDV technology can quantify energetic changes for individuals who *drink full-spectrum living water*. It can also quantify the amount of light held in a drop of water. Data on the following pages was generated using GDV technology.

The images on the opposite page (top) are of a 55-year-old woman before and 15 minutes after drinking one glass of water processed through a Fountain of Life (see page 147). Data represents the *sympathetic* nervous system which reveals information on the physical body. Note how the energetic parameters expanded significantly. This is an indication of how drinking *full-spectrum living water* can often impact the physical body.

The readings on the opposite page (bottom) are of a healthy 59-year-old man before and 20 minutes after drinking 6 oz. of water processed through a Fountain of Life. The water for this test was processed 24 hours prior to testing to reveal the ability of the water to hold its energetic capacity. The readings shown in this set are of the *parasympathetic* nervous system, which reveals information on the emotional/psychological body.

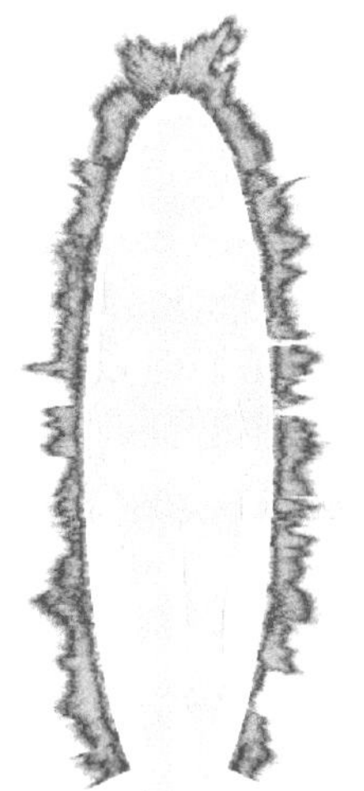

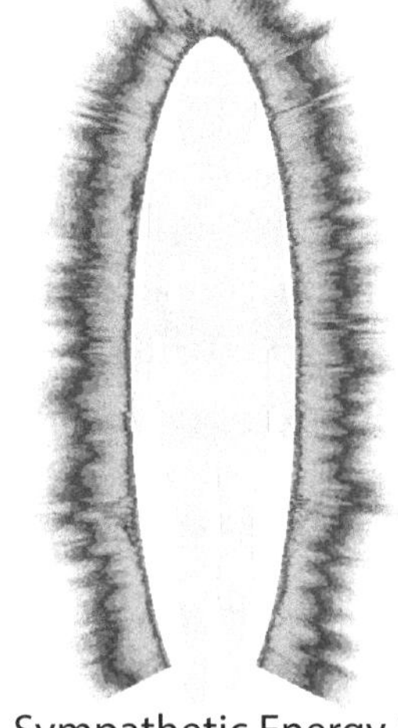

Sympathetic Energy Field BEFORE

Sympathetic Energy Field AFTER

Data provided by Dr. Sam Berne, a holistic health pracitioner who works with vision to heal mind, body and spirit. Visit www.newattention.net.

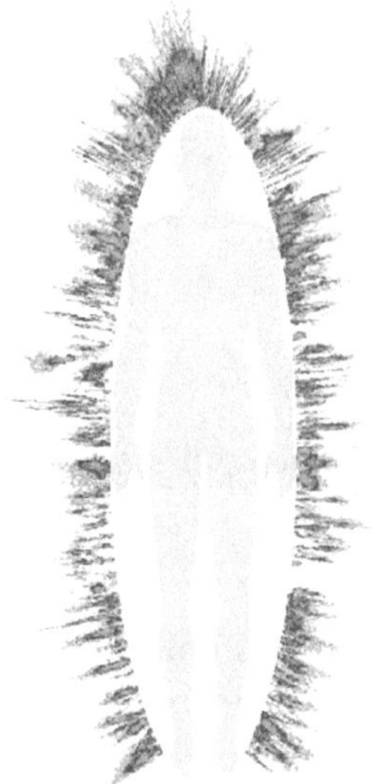

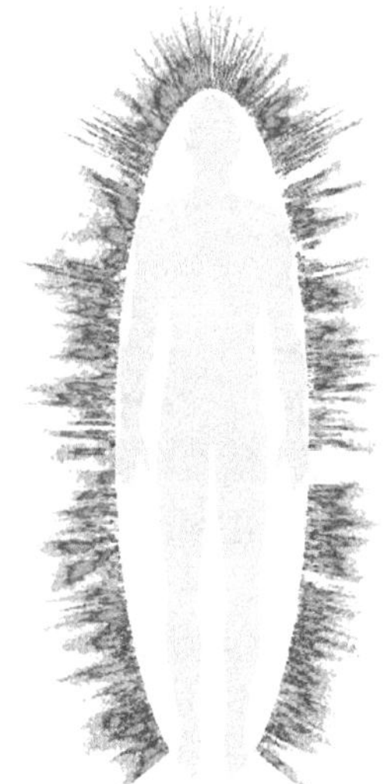

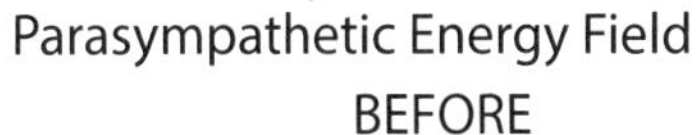

Parasympathetic Energy Field BEFORE

Parasympathetic Energy Field AFTER

Data provided by Krishna Madappa

For information on his work visit: www.krishnamadappa.com

GDV Analysis of Water

Coherent domains of structured water emit light as biophotons in a sustained reaction involving excited electrons. GDV techniques are able to quantify these emissions in a water droplet. The following *before* and *after* images of water droplets illustrate the effects of structuring techniques using several of the methods mentioned in this book. In all cases, the devices generated significant changes to the water—more evident by the data (numbers) than by the photos. Data revealed marked changes in entropy (a measure of disorder), indicating significantly greater organization of the water with all structuring treatments.

A variety of different starting water samples were used to illustrate how the quality of the initial water influenced the degree to which the water could be affected. Photos were taken by different individuals, and there is some variance (noted) between treatments.

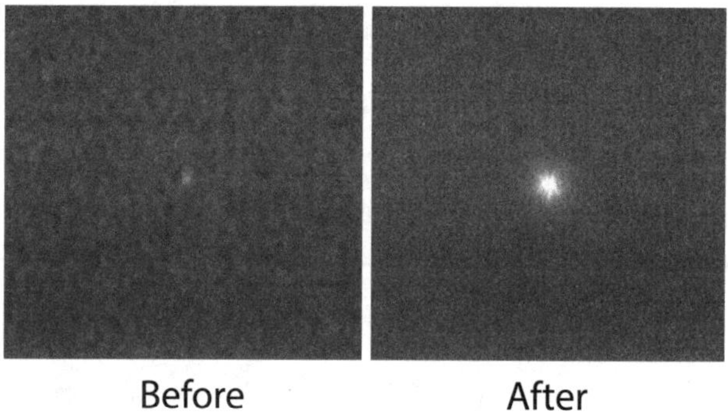

The above images show distilled water purchased at a convenience store before and after nine minutes of vortexing with the Tribest Duet Water Revitalizer (see page 144). Notice the almost complete lack of light energy in the distilled water before vortexing. (It was necessary to enlarge the first image way out of proportion just to show the speck of light in distilled water.)

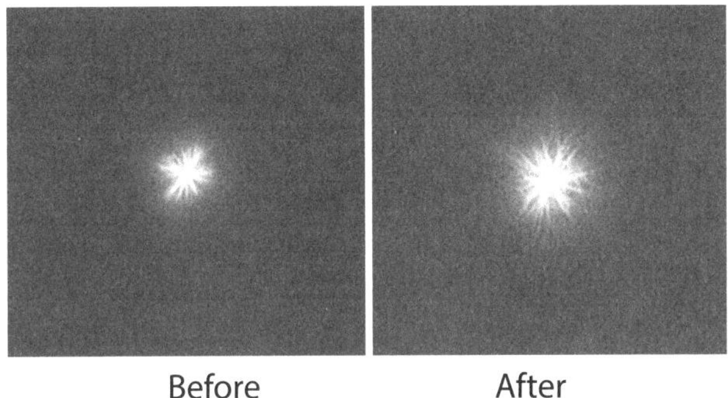

Before After

The above images show municipal water from the town of Lucy, NM, before and after treatment, using a vortexing device called the River of Life, made by Alive Water Systems (see page 225). The original water is exceptionally good to begin with, and although it is difficult to see changes, they were more apparent when the rest of the data was analyzed.

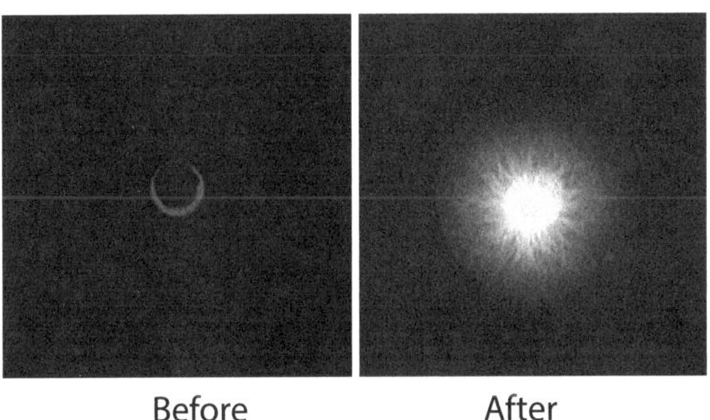

Before After

The above images show Boulder, CO, tap water before and after exposure to ANCHI Crystals. These images are proportionally larger than the previous images.

Our scientific explorations are guiding us to acknowledge the awakened presence of the fractal universe. Within this field, all systems interplay in a synchronous harmony that in Sanskrit is referred to as Anhat (stillness of presence). This presence is imprinted within us all and it appears to be awakened as we gratefully acknowledge and learn to dance with our Divine Water Partner.

– Krishna Madappa

Appendix B

The Electrolysis of Water: Water Ionizers, and Alkaline Ionized Water

The *electrolysis*, or *ionization*, of water is best known for its use in the commercial production of hydrogen and oxygen gases. The process uses electricity (direct current) to split water molecules into their constituent ions.

$$H_2O \gg H^+ + OH^-$$

Electricity then causes positively charged hydrogen ions (H^+) to migrate to the negatively charged electrode (cathode) to form hydrogen gas (H_2). Oxygen gas (O_2) is formed at the other electrode as hydroxyl ions migrate to the positively charged anode.

Alkaline ionized water

Empty water is not a good conductor of electricity, so for electrolysis to work efficiently, the water must contain minerals, or salts. In the 1930s, the Japanese began to experiment with the production of two separate streams of water containing the alkaline (OH^-) and acidic (H^+) ions resulting from electrolysis. Alkaline ionized water also carries the positively charged mineral ions (Ca^{++}, Mg^{++}, K^+, Na^+, and Fe^+) to balance the OH^- ions, and the acidic stream of water carries the negatively charged ions (CO_3^{2-}, SO_4^{2-}, Cl^-, NO_3^-, etc.).

$$H_2O + NaCl \gg Na^+ + OH^- \text{ (in solution)} + H^+ + Cl^- \text{ (in solution)}$$

Water + Salt » Alkaline water + Acidic water

(NaCl can be replaced in the above equation for any of the salts.)

Water ionizers came into vogue in Japan in the 1980s and 1990s, but by the turn of the century, conflicting information as to their long-term health benefits began to emerge.[1,2,3]

Although alkaline ionized water can be therapeutic in limited amounts for short periods of time, its strong electron potential and high pH may also be imbalancing when consumed exclusively for a longer duration. Marketers of alkaline ionized water have historically described three benefits from alkaline ionized water that are now being challenged. Each supposed benefit is discussed below:

1. Alkalization of the body from high pH water

Looking at the pH scale, a person might get the impression that anything with an alkaline pH would neutralize acids (and acids in the human body). But if that were true, then eating meat would reduce acidity in the body. It does not. Cooked meat is alkaline, yet it contributes to acidity in the tissues of the body. A lemon (one of the most acidic fruits) is well known to alkalize the body and support a healthy, balanced pH. In the case of alkaline ionized water, the "alkaline" minerals and their hydroxyl counterparts (CaOH, MgOH, KOH, NaOH, etc.) raise the pH, but they do not necessarily contribute to the neutralization of acids in the body. Why? Because the body uses bicarbonate buffers to neutralize acids[4] (see pH and buffering capacity in Chapter 5). In fact, the intake of inorganic "alkaline" minerals (especially calcium) without balancing organic acids can often do more harm than good. Inorganic minerals can end up as arterial plaque, deposits in joints, gallstones, kidney stones, bone spurs, and waste sequestered elsewhere in the body, depending on each individual's predispositions (see the section on organically complexed minerals in Chapter 5).

Although alkaline ionized water sometimes increases urine pH, urine pH is not an indicator that the body's pH is balanced. To get a more accurate picture, several other factors should be measured and interpreted together. Equally important measurements of bodily fluids are resistivity/conductivity, specific gravity, and the rH_2 value, which refers to *relative hydrogen*. Increased urine pH (a reflection of what the body is letting go of) could also mean that the body is having to get rid of excessive minerals that are not in a form it can utilize. In the long run, drinking alkaline water does not necessarily contribute to alkalization any more than eating a lemon contributes to acidosis.

2. Stronger ORP (oxidation-reduction potential) that reduces free radicals

ORP is a measure of the oxidizing, or reducing, power of a solution. When a solution has a negative ORP (measured in millivolts) it has an abundance of electrons and the potential to neutralize free radicals. Electrolysis gives water a tremendous electron potential for a short period of time. It can push the ORP to values approaching –800 mV, depending on the original water and the ionizer. That might seem desirable, but by comparison, most freshly juiced green vegetables have a more balanced ORP between –100 and +160.[5]

ORP alone is not an accurate measure of electron potential. The rH_2 value, is a more accurate representation. It measures the reducing power of a substance without the skewing effects of pH. The rH_2 scale is a logarithmic scale similar to the pH scale. Values range from 0 to 42, where 0 represents maximum reducing potential and 42 represents maximum oxidizing potential. Each whole number increase indicates a ten-fold increase in electron potential. The most commonly used equation for the calculation of rH_2 is the following:

$$rH_2 = ((ORP + 200)/30) + (2 \times pH)$$

The French professor Louis-Claude Vincent used rH_2 in the development of biological terrain assessment (BTA).

According to the principles of BTA, rH_2 values between 20 and 24 represent optimally balanced electron potential in the fluids of the human body. Similar values are acceptable for water. When you fill in the above equation with the ORP and pH values of strongly alkaline ionized water (ORP: –600; pH: 9.5), the resulting rH_2 value is 5.7. According to Vincent, that value is much too strong. Alkaline water with an ORP of –400 and a pH of 8.5 yields an rH_2 value of 10.3, which is still too strong, according to Vincent, and could cause imbalance with long-term consumption. On the other hand, a slightly reducing ORP (–50), similar to the ORP of most freshly juiced vegetables, and a more neutral pH (7.5) yields an rH_2 of 20, a value much more conducive to overall balance for drinking water.

Recent research reveals that an overabundance of reductants can be as harmful as an overabundance of oxidants.[6] Nature teaches us that balance is always better. Although alkaline ionized water may contribute to an initial reduction of free radicals, exclusive consumption of this water may upset redox balance and reduce redox signaling molecules necessary for healthy metabolic functions.[7]

3. Electrolysis creates structured water

Because water molecules are polar, they become aligned in an electric field. That fact has led some individuals to assume that electricity brings lasting molecular structure to water. Hidemitsu Hayashi (the Japanese cardiologist who did much of the original research on ionized water) proposed that electrolysis left smaller molecular "clusters" in the water.[8,9] At the time, he proposed that the smaller clusters were more easily absorbed by biological organisms.[10] However, further research indicates that small molecular clusters are associated with an overall reduction in water's structure.[11] The resulting water lacks cooperative/coherent order and is not necessarily associated with better absorption, because water molecules enter the cell single file—not in small clusters (see Chapter 9). Contrary to popularized information, the strong electrical currents used during electrolysis are disruptive of water's hydrogen-bonded network.

Alkaline water hinders digestion

Beyond the deeper look at the above widely popularized claims for ionized water, even proponents admit alkaline water may be disruptive of the digestive process. They recommend that it *not be taken with food.*

Digestion necessitates a hydrogen-rich (acidic) condition in the stomach. The pH of a baby's stomach is far more acidic than the pH of an adult's because hydrogen reserves become depleted with age. Fifty percent of people over the age of 60 are hypochlorhydric, which means they cannot produce enough stomach acid.[12] For those individuals, asking the stomach to produce more hydrochloric acid (HCl) following the ingestion of alkaline ionized water can be problematic.

Dr. Jonathan Wright has tested thousands of individuals and found that digestive problems are most often associated with *under*acidity in the stomach. For decades, vinegar has been used to aid digestion. Hydrochloric acid (HCl) supplements are often prescribed today for the same purpose. Both methods acidify the stomach and improve digestion. They provide H^+ ions.

Note: Water with bicarbonates is also "alkaline," with a high pH. However, bicarbonate water, consumed on an empty stomach actually *supports* the production of stomach acid. Bicarbonates supply the carbon dioxide required to initiate the production of HCl in the stomach. One of the best ways to support healthy digestion is with a glass of magnesium bicarbonate water an hour before meals (see the section on magnesium bicarbonate water in Chapter 5).

Benefits from alkaline ionized water are a result of the hydrogen

Although the initial benefits often reported by those who consume alkaline ionized water are not likely a result of pH or changes in the structure of the water, they can be attributed to the presence of molecular hydrogen in the water.[13,14] That was the conclusion Dr. Hayashi came to after years of work with ionized water in Japan. In fact, he eventually withdrew his support of water ionizers and developed his own method for enriching water with hydrogen. His hydrogen stick was one of the first alternative methods of producing hydrogen-rich water.

Even though some of the hydrogen is discarded in the acidic portion of ionized water, molecular hydrogen accumulates in the alkaline water. The amount is typically between 0.1 and 0.7 ppm, depending on the minerals in the original water, the flow rate, and the age/cleanliness of the electrodes. New ionizers are being developed to optimize the levels of hydrogen—especially in light of the emerging benefits of hydrogen-rich water (see Chapter 5). These newer machines will undoubtedly favor a more neutral pH and a higher concentration of molecular hydrogen in the resulting water. However, the process is still an electrolytic process that has other effects on the resulting water.

Electrolysis increases the amount of deuterium in the water

One subject that promoters of electrolysis have not considered carefully is the fact that the resulting water has a greater concentration of deuterium (a hydrogen isotope with a neutron in the nucleus of the atom). Deuterium increases hydrogen-bond strength and requires more energy to break.[15] Thus, during electrolysis as hydrogen bonds are broken to release hydrogen and oxygen, the stronger deuterium bonds remain, leaving the resulting water with a greater concentration of deuterium. Even a small increase in deuterium can have an aging effect on those who consume the water[16,17] (see Chapter 15).

Electrolysis summarized

As discussed previously, the strong electrical current used for electrolysis disrupts the hydrogen-bonded water network. In this state, its ability to carry the finely tuned resonances essential for organic life is compromised. With a "too powerful" electrical potential (ORP), and with an abundance of imbalanced and inorganic minerals, ionized water is unnatural. Increased amounts of deuterium further imbalance the water. The long-term consequences are not yet fully understood.

Studies with rats (whose entire life cycle can be studied in a short period of time) reveal potential problems with long-term, exclusive consumption of alkaline ionized water. These studies indicate that this water may lead to cardiovascular problems.[18,19,20] It may also contribute to arterial plaque, kidney stress, and digestive problems—among other things.

Nature is about balance—not extremes. When the energy used during any process destroys water's balance, there is a potentially negative effect on its life force and a potentially negative outcome for any life form that consumes it. In traditional Chinese medicine, taking strong "yang" herbs and other substances (like alkaline ionized water) for too long, overheats and burns the body out. Ultimately, these practices contribute to aging. Although alkaline ionized water has been shown to be therapeutic for short periods of time (most likely as a result of its molecular hydrogen content), the authors have found alkaline ionized water to be harsh and aggressive. They recommend that it not be consumed on a regular or long-term basis.

References (Appendix B)

1 Watanabe, T., Kishikawa Y., and Shirai W. (1997). Influence of Alkaline Ionized Water on Rat Erythrocyte Hexokinase Activity and Myocardium, *Journal of Toxicological Science*, 22(2), pp. 141-52.

2 Watanabe, T. and Kishikawa Y. (1998). Degradation of Myocardiac Myosin and Creatine Kinase in Rats Given Alkaline Ionized Water, *Journal of Veterinary Medical Science*, 60(2), pp. 245-50.

3 Watanabe, T., Shirai, W., Pan, I., Fukuda, Y., Murasugi, E., Sato, T., Kamata, H., and Uwatoko, K. (1998). Histopathological Influence of Alkaline Ionized Water on Myocardial Muscle of Mother Rats, *Journal of Toxicological Science*, 23(5), pp. 411-17.

4 Frassetto, L. and Sebastian, A. (1996). Age and Systemic Acid-Base Equilibrium: Analysis of Published Data, *Journal of Gerontology: Biological Sciences*, 51A(1), pp. B91-B99. Available online: http://biomedgerontology.oxfordjournals.org/content/51A/1/B91.full.pdf

5 Pinto, V. (2001). Extraction and Storage of Raw Juices While Retaining Nutrient Quality. Available online: http://rawpaleodiet.vpinf.com/raw-juice-refrig-storage-summary-overview.html

6 Samuelson, G. (2009). *The Science of Healing Revealed: New Insights into Redox Signaling*, pp. 27-32 and 50-51.

7 Samuelson, G. (2009). *The Science of Healing Revealed: New Insights into Redox Signaling*, p. 30.

8 Hayashi, H. (1996). Microwater: The Natural Solution. Water Institute: Tokyo, Japan.

9 Hayashi, H. Understanding Alkaline "Ionized" Water. Available online: http://www.heartspring.net/water_science.html

10 Hayashi, H. Understanding Alkaline "Ionized" Water. Available online: http://www.heartspring.net/water_science.html

11 Chaplin, M. LSBU Website: http://www1.lsbu.ac.uk/water/magnetic_electric_effects.html

12 Kitchen, J. (2001). Hypochlorhydria: A Review Part I. *Townsend Letters for Doctors and Patients.*

13 Fujita, R., Tanaka, Y., Saihara, Y., Yamakita, M., Ando, D., and Koyama, K. (2011). Effect of Molecular Hydrogen Saturated Alkaline Electrolyzed Water on Disuse Muscle Atrophy in Gastrocnemius Muscle, *Journal of Physiological Anthropology*, 30(1), pp. 195-201.

14 Xue, J., Shang, G., Tanaka, Y., Saihara, Y., Hou, L., Velasquez, N., Liu, W., and Lu, Y. (2014). Dose-dependent Inhibition of Gastric Injury by Hydrogen in Alkaline Electrolyzed Drinking Water, *BMC Complementary and Alternative Medicine*, 14(1), p. 81.

15 Goodall, K. (2003). Preliminary Analysis of Deuterium's Role in DNA Degradation, *Anti-aging Medical Newsletter*, July 22, pp. 7-31.

16 Griffiths, T. (1975). The Possible Roles of Deuterium in the Initiation and Propagation of Aging and Other Biological Mechanisms and Processes. Proceedings of the Second International Conference on Stable Isotopes. October 20-23, Oak Brook, Illinois.

17 Somlyai, G. (2001). *Defeating Cancer: The Biological Effects of Deuterium Depletion*, 1st Books Library.

18 Watanabe, T., Kishikawa Y., and Shirai W. (1997). Influence of Alkaline Ionized Water on Rat Erythrocyte Hexokinase Activity and Myocardium, *Journal of Toxicological Science*, 22(2), pp. 141-52.

19 Watanabe, T. and Kishikawa Y. (1998). Degradation of Myocardiac Myosin and Creatine Kinase in Rats Given Alkaline Ionized Water, *Journal of Veterinary Medical Science*, 60(2), pp. 245-50.

20 Watanabe, T., Shirai, W., Pan, I., Fukuda, Y., Murasugi, E., Sato, T., Kamata, H., and Uwatoko, K. (1998). Histopathological Influence of Alkaline Ionized Water on Myocardial Muscle of Mother Rats, *Journal of Toxicological Science*, 23(5), pp. 411-17.

The consumption of deuterium-depleted water gradually reduces the deuterium concentration in an organism. This supports normal DNA replication and repair and the production of ATP. At the same time, a number of health parameters may improve.

Appendix C

Deuterium-depleted Water

Deuterium is an isotope of hydrogen. Because it contains a proton *and* a neutron in the nucleus of the atom, the mass of a single deuterium atom is twice the mass of a hydrogen atom. In some ways, deuterium (D) could be considered an independent element because it behaves differently from hydrogen in many chemical reactions (the isotope effect).

When deuterium combines with oxygen, the resulting water is referred to as deuterium oxide, or "heavy" water (D_2O). Semiheavy water results when one atom of hydrogen and one atom of deuterium combine with oxygen (HDO). Water made with deuterium tastes and looks the same but has distinguishing characteristics. For example:

- Water made with deuterium is bonded more strongly; it requires a greater amount of energy to boil. (Normal water boils at 100° C; heavy water boils at 101.4° C.)
- Water made with deuterium freezes first. (Normal water freezes at 0° C; heavy water freezes at 3.8° C.)
- Ice made with D_2O does not float—it sinks.

Gilbert Lewis was the first to produce a pure sample of heavy water in the 1930s. He accurately predicted its toxic effects on living organisms. His experiments showed that while tobacco seeds placed in typical water sprouted over the course of two weeks, those placed in D_2O did not sprout at all. Tobacco seeds in a 50% D_2O/H_2O solution sprouted—but slowly. Subsequent experiments with cell cultures and living organisms revealed that increasing the deuterium concentration could disturb normal cell function. Higher concentrations were lethal. Although this was interesting, most scientists considered deuterium to be so scarce that the study of its biological effects were ignored until the 1990s.

Is deuterium really scarce?

On average, one out of every 6,400 atoms of hydrogen is deuterium—the equivalent of one or two drops in a quart of water (about 150 ppm). That might seem like a small proportion until you consider the amount of hydrogen in a living organism. One out of every 6,400 atoms translates to a mass of deuterium that is five times greater than the mass of calcium in human blood. As scientists began to look at deuterium from this perspective, they became more interested in its biological significance. According to clinical work, even a seemingly small reduction in deuterium can have significant biological effects.

Biological effects of deuterium-depleted water

Because water is a major source of hydrogen and because most living organisms are 50% to 75% water, the effects of deuterium-depleted water (DDW) are no longer being ignored. The biological effects of DDW, often referred to as "light" water, have been highlighted in dozens of studies conducted on plants, animals, and humans. Experiments on plants show a general increase in growth and productivity. Experiments with chickens reveal a decrease in mortality and an increase in egg production. Experiments with mice show increased life expectancy and an increase in sexual activity. Over ten years of research and clinical trials on humans in Hungary by Dr. Gábor Somlyai and colleagues have identified positive effects of DDW in treating diabetes and many types of cancer. Not only does the consumption of DDW have a positive effect on the progression of many diseases and metabolic disorders (including hypertension, obesity, and diabetes), according to research conducted in Russia, it also positively affects energy production and immune function. A growing body of evidence shows DDW stimulates skin regeneration, reduces inflammation, and slows aging. It also mitigates the damaging effects of radiation.

Deuterium and DNA

The most widely accepted theory on aging proposes that the aging process is correlated with a gradual accumulation of errors in DNA. According to Kirk Goodall, a senior member of the technical staff with NASA, the number of irreversible errors in the DNA sequence is directly influenced by deuterium. It affects the shape of molecules, including the shape of enzymes, many of which are involved in DNA synthesis and repair. The presence of deuterium in those enzymes is thought to slow DNA replication, cause errors in transcription, and hinder DNA repair. The lower the deuterium concentration, the lower the number of errors.

Hydrogen bonding plays a major role in water's structure and also in the structure of DNA. In fact, it is the organized structure of water that supports the helical shape of DNA. The neutron in deuterium may create a "pucker" in the liquid crystalline matrix, contributing to an overall destabilization of DNA. Deuterium bonds are stronger than normal hydrogen bonds; they require more energy to break. This may stiffen proteins, another factor related to aging.

Deuterium and proton movement

Deuterium also appears to interfere with normal proton movement in proton channels, including the channels involved in ATP production. The presence of deuterium is thought to cause the mechanism to "stutter." With normal concentrations of deuterium, the "stutter" happens about once every 15 seconds in each proton channel. When multiplied by the millions of proton channels in a living organism, this approaches significant proportions. Given these effects (and others), deuterium is now thought to play an important role in the progression of disease and aging.

Natural deuterium depletion

Evaporation favors hydrogen over the heavier deuterium atom. In other words, water vapor is lower in deuterium. (Deuterium evaporates last and condenses first.) In areas where there is greater evaporation (equator and deserts) the deuterium content of the surface water is high. Polar and mountainous regions have lower concentrations. Natural deuterium concentration varies depending on a number of factors:

- Temperature/season—Water in cold climates contains less deuterium than water in warmer climates. Winter precipitation contains less deuterium than summer precipitation.
- Water source (fresh vs. ocean)—Oceans contain more deuterium than freshwater. The deuterium concentration in the Atlantic and Pacific remains fairly constant at 156 ppm. Polar oceans have a much lower concentration.
- Distance from coastline—Surface water along western coastlines contains more deuterium than inland areas because the heavier water precipitates first.
- Altitude—Water at high altitudes has less deuterium because heavier water results in rain, leaving snow with less deuterium.
- Distance from the equator—Equatorial waters contain more deuterium than water at the poles. Water from Antarctic ice measures 90 ppm deuterium, and water beneath the Sahara Desert measures 180 ppm deuterium.

Why drink deuterium-depleted water?

Organisms in various parts of the world tend to have deuterium concentrations comparable to the water in the area. However, both plants and animals maintain a lower deuterium concentration than the surrounding surface water. This indicates that organisms have a preference for a deuterium-depleted status. In fact, the water produced in the body (referred to as metabolic water) is deuterium depleted.

The consumption of DDW gradually reduces the deuterium concentration in an organism. This supports normal DNA replication and repair and the production of ATP. At the same time, a number of health parameters may improve. This was shown during case studies conducted by the authors of *Dancing with Water* (results available on the Internet). Dr. Somlyai's work in Hungary revealed that healthy cells respond well to reduced amounts of deuterium in water. However, cells with chromosomal mutations (cancer) are more sensitive to deuterium depletion. Cancer cells, particularly tumor cells, cannot adapt quickly, resulting in tumor regression without side effects. Dr. Somlyia and his colleagues also studied DDW for metabolic disorders with favorable results.

In the authors' experience, when DDW is structured and conditioned using the techniques discussed in this book, less is required to produce potent health effects. As little as one-half cup per day of DDW with a low concentration (25 ppm) can provide measureable health benefits.

Suggested reading:

1. Somlyai, G. (2002). *Defeating Cancer! The Biological Effect of Deuterium Depletion.*

2. International Congress on Deuterium Depletion Website: http://www.deuteriumdepletion.com/

3. Boros, L. D'Agostino,D. Katz, H. Roth, J. Meuillet, E. and Somlyai G. (2016). Submolecular regulation of cell transformation by deuterium depleting water exchange reactions in the tricarboxylic acid substrate cycle. *Medical Hypotheses.* 87(1), pp. 69-74.

Index

About The Authors

MJ Pangman is a writer and natural scientist who has authored or ghost-written a number of books for doctors, naturopaths and inventors. In the 1990s she became fascinated with water as a vehicle for the transmission of energy and information, and as a conscious participant in the "dance of life." Her research and writing of *Dancing with Water* has allowed her to step into a new relationship with the Earth and with the natural forces that constitute the "Web of Life." She is committed to helping others gain a similar appreciation for water's multi-faceted role in life on this planet. She looks forward to the day when the human race will honor and live in harmonious balance with Mother Earth.

Melanie Martin is an intuitive and a healer whose earliest memories are of connecting with the Earth. She remembers being at the ocean as a 3-year-old and of understanding the life force there. She has spent many hours merging with water in a way similar to how Victor Schauberger described merging with water's consciousness in the Austrian forest. To Melanie, water is a kindred spirit. Working on *Dancing with Water,* she has been intrigued with the way water carries vibrations and patterns and by how these are enhanced with vortices, magnetic fields, thoughts, and other forms of subtle energy. She can't wait to have others join her in the "dance" she has always danced with water.